PEDRO GADANHO

ACTAR

Dedicated to my children, Catarina and Miguel

Ten essays zooming into the ways the environmental crisis impacts architecture and its practice, while probing into historical background and some critical reflections on the discipline's current *status quo*.

"The most startling aspect of this story is just how much these people knew, and how unable they were to act upon what they knew. Knowledge did not translate into power."

—Naomi Oreskes, in *The Collapse of Western Civilization*, 2014.

Climax Change!   Pedro Gadanho

CLIMAX CHANGE!

HOW ARCHITECTURE MUST TRANSFORM IN THE AGE OF ECOLOGICAL EMERGENCY

Stop Building: A Prelude

More than a decade ago, shortly after the 2008 financial crisis, I wrote a small, polemical pamphlet defending that architects should be the first to tell society a difficult truth: we should stop building anew. Architects should resist a natural temptation to embrace any opportunity to erect a new building, and they should warn society of the hidden future costs of building frenzies like the one that had just helped trigger the great recession.

With their specific knowledge of the impacts of any new additions to the built environment, architects should overcome a selfish want to build from scratch and let people know that there are viable alternatives in recycling existing constructions. They too should join emerging pleas for the three Rs. As designers, planners and creators they should be the first to assume responsibility in helping the construction industry rethink how to reduce, reuse and recycle.

See Pedro Gadanho, 2009 "Parar de Construir, ou o Regresso à Cidade", in José Manuel das Neves (ed.), Living City, True Team, Lisbon, 2009

More than the expression of an urging environmental concern, that short essay was a reflection on the paradoxes of an ever-growing building supply in the face of ever scarcer natural resources. Although alluding to sea level rise and ecological distress, the piece rather underlined the need to avoid the economic pressure to embrace new construction in virgin territory. It stressed the urgent need to move forward to practices of rehabilitation and renovation.

Needless to say, the arguments I advanced were coming from a Eurocentric perspective, in a context in which populations were starting to shrink. A situation in which the existing building stock should be able to cover existing and future demand. Even where new needs would arise, conversion of available buildings could provide creative solutions. Yet, anytime I described such notions, immediate protests rose about housing shortages, the urges of developing economies or again the permanent want of new building typologies for novel uses.

Eventually, I gave up on the discussion.

It was only when my awareness of the dimension of our climate crisis rose to new levels that, a couple of years ago, the idea resonated again in my mind. With the exception of one or another thought experiments along the way, it was also only recently that the idea resurfaced in architectural discourse, if still very discreetly.

So, when I finally decided to take on the project of writing about climate change and its impacts on architectural practice, this seemed like a subject worth revisiting. As calls arose for fossil fuels to be left in the ground, it seemed appropriate to consider that new buildings could also be left on paper. In Russia, in the eighties, a wave of 'paper architecture' was a sign of political resistance to a corrupt system. Elsewhere, many contributions to architectural discourse have remained influential without ever leaving the paper on which they were ideated. Fictional architecture at some point grew into a real possibility to inform and enrichen the culture of architecture.

In the face of a growing — and necessary — environmental consciousness, perhaps it was again time for architects to let go of the peer pressure to achieve recognition only through the production of new buildings.

This would imply, first of all, that architects shifted the importance they put on new construction — as a hallmark of authorship, disciplinary legitimation and even the major avenue for architectural innovation. Such a shift would certainly help conceive how our society can stop new construction. It could actually be a good illustration of one of those 'climax changes' that this book calls for.

If architecture must undergo a paradigm shift so as to join collective efforts in tackling an encroaching environmental crisis, then the symbolic acts from which we architects get our kicks must also surely change.

A sort of unique orgasmic realization still seems only possible when delivering a shiny new building. So, perhaps we must try and shift the locus of our typical collective architectural desire. As ecological awareness becomes more asphyxiating, we must try and redirect our lust somewhere else, and thus allow ourselves to argue for halting the ongoing rape of virgin land.

As we accept urban densification as one possible strategy for carbon dioxide reduction, we architects should definitely start to conceptualize the trade-offs that would allow us to tell society to ban building. But only a full consideration of implications, options, costs and benefits will allow us to properly advocate that new construction beyond the boundaries of urban nodes should stop as soon as possible.

In his book *A Country of Cities*, architect Vishaan Chakrabarti did not go so as far as to suggest that we should halt new buildings altogether. Nonetheless, he provided abundant signs that behaviors in the construction world, as in everyone's lifestyles, must per force undergo some kind of collective transformation. As he wrote, "it is now abundantly clear that we use more natural resources than the world can replenish." With the demonstrated small impact of "feel-good actions," and with the full perception of how unsustainable our practices are, we should "address resource usage at a global scale"[1] — including notably how we refurbish and make our cities more efficient. Chrakabati goes to lengths to reveal how the densest cities are proven to "produce the lowest greenhouse-gas emissions per capita." So, this is the first context in which a halt on new construction must be considered.

1. See Vishaan Chakrabarti, *A Country of Cities*, in Dan Willis, William W. Braham, Katsuhiko Muramoto, Daniel A. Barber (eds.), *Energy Accounts: Architectural Representations of Energy, Climate, and the Future*, Routledge, New York: 2016.

When calling for a ban on new buildings in non-urban territory, one can actually imagine a regulatory pathway to achieve a built environment with less of a carbon footprint. And one would not be calling for the eradication of the construction industry — as much as many now tentatively ask for the extinction of fossil fuel industries. The lure for such a call is certainly there. When one knows cement alone is responsible for 8% of global carbon dioxide emissions, one may be enticed to plea for its exclusion once and for all.

But, no, we should rather think in terms of incentives. We should rather encourage that, while greening its practices, the construction industry, as society in general, must redirect its efforts to densifying and renewing those cities that are still attracting newcomers from their surroundings. So, perhaps a few new buildings will be allowed, but only when they would be contributing to urban densification — and never when expanding the actual foothold of existing cities on their less built surroundings.

In my searches, I did find one other lonely advocate of the need to stop building. Writing fot the HuffPost in 2013, architect Lance Hosey hypothesized how a ban on building would work.[2] He was pulled to that thought experiment by the news of yet another psychological climate barrier broken at the time. This was the first-time scientists registered 400 parts of carbon dioxide per million in the Earth's atmosphere, surpassing records of the last 800,000 years, much before human civilization appeared.

As we now read on an everyday basis, Hosey underlined estimates pointing to the construction industry as the culprit for nearly 40% of carbon dioxide emissions. He pointed out that LEED certification strategies would not be enough to reduce the footprint of building's embodied carbon in a significant manner. While such strategies had helped cut emissions, a report by the Preservation Green Lab[3] stated that it could take "up to eighty years for the greater energy efficiency of a LEED-certified building to overcome the harmful climate effects created during the building's construction." This would, of course, represent a blunt failure to meet current demands to achieve zero emissions by 2050. The alternative, then, could be to temporarily ban new construction.

The arguments for such a "shock proposal" reminded me of what prompted my own thought experiment a few years earlier. In the United States, too, the US Census Bureau pointed to a gross vacancy rate of buildings at around 14%. And, as was then reported in Time magazine, the Metropolitan Institute at Virginia Tech forecasted that by 2025 there would be "a surplus of 22 million large-lot homes."[4]

One could counter that this represents a problem of uneven allocation of resources. Seizing a famous quote from science fiction writer William Gibson, the future stock of all the buildings we could ever need was already here. The only problem was that it was unequally distributed.

2. See Lance Hosey. "Stop Building. Now." 14 Nov 2014. The HuffPost, Accessed Feb 27 2020. < https://www.huffpost.com/entry/stop-building-now_b_4194532>

3. See "Older, Smaller Better" report, published online 11 March 2016. SavingPlaces.org. Accessed 28 Feb 2020. https://forum.savingplaces.org/connect/community-home/librarydocuments/viewdocument?DocumentKey=83ebde9b-8a23-458c-a70f-c66b46b6f714

4. See Brian Walsh, "Recycling the Suburbs," 12 March 2009. Time Magazine. Accessed 28 Feb 2020. http://content.time.com/time/specials/packages/article/ 0,28804,1884779_1884782_1884756,00.html

But here is the interesting thing: one can actually include this fact in any equation of how to go about an effective ban on new construction. For Hosey, a temporary, nationwide moratorium could be aimed at the "suburban greenfield sites" whose growth "often exacerbates sprawl, increases emissions, and destroys habitats." This would force the construction industry to "turn its attention to improving the nearly 300 billion square feet of existing building stock." The only allowed exceptions would be "net zero" constructions on brownfield or reclaimed infill sites, with preference given to "sites where soils are contaminated and remediation would be required."

While these sound like ecologically sensible ideas, one could expect the usual outcry about the unacceptable disruption to a much-needed economic activity. So, I would add some strong economic arguments to that potential ban policy package.

Consider this.

Like in carbon offsets, or affordable housing compensation schemes, entirely new construction in urban areas could also be allowed, but only if developers would commit to certain trade-offs. They could, for instance, take on the job of buying, demolishing and rewilding land plots of similar dimension in adjacent non-urban areas. They could also be required to include at least 20% of recycled materials from demolition sites in any new building projects. This, as any other regulations, would drive construction prices up. Yet, from the point of view of environmental economics, the value of community and ecological benefits could be proven to offset the marginal cost surges.

Firstly, economic and ecologic asymmetries would be slowly corrected: while cities would be getting denser to respond to higher building demands, areas with less or zero demand would be returned to natural eco-systems or regenerative agriculture. Along with traditional urban development planning, one should expect demolition planning strategies to be born at the regional level. Thoughtful consideration should be given to what to demolish first, attending to community needs and to what sites were effectively being abandoned. By analogy, remember how many blocks in the city of Detroit were emptied out and progressively turned into restorative green spaces.

Secondly, there would be similarities to how market rules determine when it compensates to extract fossil fuels — or just leave them in the ground. Demand and prices would determine when it would really pay back to invest in a new building, rather than retrofitting an existing one. By inverting the current tendency, as one must, it could soon prove economically more attractive to recycle rather than building anew. Higher prices would create a market constraint to the temptation of producing excessive emissions. Economics mechanisms would provide the driver for a beneficial contribution to the new environmental balance, which we must per force achieve within the next few decades.

Now, one can ask a poignant question. This book is supposed to be about how future practices of architecture will be impacted by the climate crisis. So, what does such a regulatory thinking have to do with architecture?

Truth is that, as we will see in other chapters, whether architects like it or not, architecture practices are highly dependent on market demands and the workings of the construction industry. But that does not mean that architecture may not as well influence industry practices. Design thinking may well extend beyond its traditional narrow confines.

Once, architectural discourse made early modernism attractive to architects. But that same discourse also eventually led builders and developers to adopt a novel architectural language, along with the new materials and techniques it brought along. With the climate crisis, we are faced with the need for similar, if not bigger, radical change in the logic and practice of architecture. Industrialization brought about a new architectural aesthetics. The environmental emergency — and the *Great Transition* that it implies — should have a similar, if not bigger, effect.

As architects become more knowledgeable in conceiving net-zero emissions buildings, or more experienced in creatively reducing costs in retrofitting or recycling operations, they too can contribute to the changes that are being required at wider scales. As they make ecological needs central to their form-finding, they can become important actors in a radical change that goes well beyond their limited field of action.

As Hosey suggested, if "any architect or builder not interested in renovation or preservation suddenly would become an expert in high-performance design, the intelligence and innovation of the entire industry would skyrocket." That is, architects must start re-directing their creative power, as well as their lobbying capacities, to support urgent transformations. Then, they too can be a powerful part of the shift required of any sector that bears some responsibility for impacts of the built environment on sustainable development.

With growing signs of a massive environmental crisis lurking around the corner, architects and planners must actively reassess how their specific knowledge brings increased responsibility. They are particularly well-positioned to alert society to the need for drastic changes in the way we transform the built environment. And this is not only the case when considering a potential ban on new buildings. It implies wider considerations of how the climate crisis will affect populations and territories.

As noted in a recent Washington Post article, mass migrations due to sea-level rise or increased droughts will result in massive pressure on urban settlements that were not preparing for unexpected population influxes. However, as Steven Bingler and Martin C. Pedersen point out, "few of our politicians will 'go there' yet, because their planning for the future extends precisely as far as the next election." That means that "it's time for architects and planners to sound the alarm."[5]

5. See Steven Bingler and Martin C. Pedersen, "Planners talk about resilience in the face of climate change. We need to start using a different R word." 17 Feb 2020. Washington Post, Accessed 1 Mar 2020. <https://www.washingtonpost.com/opinions/2020/02/17/climate-hope-best-prepare-worst/>

Sounding the alarm implies that, as we rebuild, new ecological balances must be introduced in the planning equation. As retreat substitutes for resilience, some cities will lose population, while others will be forced to accommodate them. And then, we surely must take into consideration how demolition and rewilding will have to compensate for the densification of safer urban nodes.

In a way, the apparently crazy idea of banning new construction is only part of the scenario thinking that we must engage in, if we want to embrace the change that is coming to our architectural and planning practices. It is but one of the thought exercises we must embrace if we want to reshape the aspirations that drive those same practices. As more recent contributions to the architectural discourse have started to revisit, the bewildering notion of "no new buildings" can be just another option in the wide menu of choices we have to consider when facing a global ecological emergency.[6] Stopping new construction, except if in very well-defined conditions, is but one of the thousand pathways that we must address and thoroughly consider when the environmental crisis forces an authentic paradigm shift in architectural practices.

6. See Thomas de Monchaux, "A New Idea in Architecture? No New Buildings." 30 December 2019. Metropolis magazine. Assessed 14 August 2020. https://www.metropolismag.com/architecture/new-idea-architecture-no-new-buildings/pic/66291/

In that vein, it is important to reassure those architects who have been resisting climate-related change as a menace to their beloved design beliefs. For their sake, it is certainly important to clarify that considering these pathways for change may actually represent an opportunity for architecture to gain a new breadth. It is good to clarify that exploring the avenues that the climate crisis imposes on us does not necessarily collide with the most cherished, essential principles guiding the discipline of architecture.

Scenarios such as that of banning new construction apparently go against the grain of architectural desire. And yet, we may come to realize that these scenarios trigger new design problems through which architects can uncover new, unknown pleasures — but also, as early modernists knew well, rediscover the creative climaxes of addressing higher transformative goals. As this book will propose, anticipating and dissecting a paradigm shift in architectural practice does not represent a call to abandon aesthetics or form-making as one of the essential aspects of architectural discourse.

For sure, these speculative operations hint at the scope and breadth of shifts that are required to start major change. Embracing such shifts may include a new architectural language that integrates and naturalizes urgent ecological concerns. It may include the ability to conciliate the inner needs of architecture as a creative and autonomous endeavour with new, urgent expressions of our shifting ecological *zeitgeist*. As a strong believer in the creative power of design thinking to tackle wider human problems, I would suggest that the advancement of spatial disciplines now lies at this precise intersection.

At this moment, we scarcely need to invent another obscure theoretical excuse to make architecture originality leap forward. We really do not need another representation technique to makes us feel excited about how architectural may look sexy to outsiders. In a moment of growing anxiety, the next formal or spatial cheap trick is not what will make the profession relevant for society. Indeed, we again need a manifesto for change. We need an outline of what it takes to effectively embrace a paradigm shift in practice.

Looking at the scale of transformations implied by the coming environmental crisis, we need to revive the spirit that once drove the avant-gardes. We need to redirect that creative energy at the issues that are now proving to be inescapable — as war, rapid urbanization or technological breakthroughs were in the past. And this points to the urgency of a new sort of *unified theory* of architecture, in which we can encompass and make sense of both the obstacles and the potential pathways so as to embrace the change that lies ahead of us.

1. What Ever Happened to the Environmental Avant-garde?

In early 2020, right before the corona virus pandemic numbed us with full force, my attention got caught by a modest project of an off-grid, zero emissions cabin somewhere in Chile.

While musing about stopping all new construction, I thought that if you do have to build something in uncharted territory, then at least let it be a self-sufficient structure that leaves a minimal footprint on its surroundings. And such cabins, with their implicit desire for an exclusion from the perils, pressures and viruses of modern life, had regularly been around as a promise of environmental fulfilment. Ever since Marc-Antoine Laugier assimilated the primitive hut into architectural discourse, the isolated, somewhat rustic cabin firmly sunk in architects' imagination as the ultimate alternative to more complex, historical-heavy conceptions of architecture.

Laugier's *Essay on Architecture* came out in 1755. It was a primer of Enlightenment thinking, anticipating the French Revolution by a few decades. As humanity drifted away from some vague notion of paradise lost, the primal cabin continued to present a promise of reconnection to architecture's opposite, Nature herself. Even Le Corbusier, that great propeller of modern urban transformation, crowned his professional endeavours with the construction of a small sea-side cabin. This was still idealised as an architectural microcosm, but one to which he could eventually retire, finally freed from excessive formal attire.

Essay on Architecture
by Laugier

So, that little cabin in a remote landscape of Chile didn't catch my attention because it represented the nth attempt at human escape from the urban environment. It didn't surprise me because it was again enmeshed in the dream of energetic self-sufficiency in the middle of nowhere.

No.

What caught my attention in this contemporary hut was the fact that, although it was appearing

Edifice by Marc Thorpe.
Photo by Marco Petrini

in architectural media, there were no architects involved in its conception. What caught my attention was the fact that, with a climate emergency emerging all around us, the irresistible off-grid cabin was suddenly being kidnapped from the architectural imaginary. The new primitive hut was being snapped into popular DYI culture right from under architects' noses.

Here was a housing project that ticked all the musts in a menu of how to tackle the climate emergency through architectural design. Here was a do-it-yourself kit full of all the requirements needed to address the *Great Transition* into a new energy regime. From rainwater collection and reuse, to solar power, to the ability to "incorporate power-generating devices such as micro water turbines," the little cabin had it all. And while the dwelling was rented out to tourists, the proceedings would go towards the wildlife preserve where it is set. A kind of manual of good architectural practices for the times ahead — even if at small scale, and not really yet accessible to every urban dweller on the globe.

Yet, the point was that there were no architects in sight. No notice of architects involved, even though beyond its forward incorporation of green technologies, the cabin is actually aesthetically adequate. Which inevitably makes one wonder where architects have been investing their creativity, especially over these years in which the ecological crisis has become rampant. Why are architects not involved in making this offer of a zero-emissions hut, or for that matter, a full zero-emission housing package, also an amazing architectural statement?

In fact, around the period in which Zero-Cabin made its way into architectural media, other such cabins appeared that carried the preferred minimalist signature of current contemporary architecture. One was designer and architect Marc Thorpe's dwelling in upstate New York, which apparently took inspiration from an individualist appreciation of nature found in Henry David Thoreau and Ralph Waldo Emerson.[7] Another was *The Hut*, by Ohio-based practice Midland Architecture. In this case, the treehouse-like dwelling was driven by a deep emotional connection to a family-owned plot.[8]

7. See https://www.dezeen.com/2020/02/08/edifice-cabin-marc-thorpe-upstate-new-york/

8. See https://www.dezeen.com/2019/03/30/hut-midland-architecture-ohio-off-grid/

Both these designs were justly deemed "sustainable" and carried all the off-grid requirements of the day. Yet, rather than an imperative to explore an architectural language determined by new needs, they bluntly confirmed the tendency to abide by an aesthetic establishment determined by a modernist formal legacy. Clearly, their beautifully composed volumes speak of everything comfortable and tidy but an ecological emergency on the go.

The fact that even this small environmental commitment was not happening at a greater scale and impact — and where it did happen, it served only to assert a dominant aesthetics devoid of any external sign of ecologically driven intentions — triggered yet another doubt: whatever had happened to the environmentally inclined architectural avant-garde? Had it got tired of unsuccessful, discarded experiments? Had it lost its flair for innovative takes on green construction? Had the ecological avant-garde been finally and ultimately subsumed by the full-capitalist logic of the neomodernist mainstream? In an age in which the environmental crisis *became* the avant-garde, it struck me that such a tasty topic

ZeroCabin, Chile by a multidisciplinary team (Felipe Lüer, Hector Becker, Luis Valladares, Andrés Lüer, Oscar Villalon, Ian Burbulis)

for architectural innovation — both in technological and aesthetic terms — was mostly being left for engineering enthusiasts.

Truth be told, however, there was a time in which architects had been ahead of the curve when it came to ecological upheaval. There was a time, around the 1970s, in which a few radical architects were out there with a few inventors, activists and early environmentalists, responding to the early signs of a budding ecological crisis. There was also an earlier time, around the 1950s, in which architects — who were far from imagining to be an avant-garde — allied with scientists to do research on solar-oriented architecture, as Daniel A. Barber recounts in his book *A House in the Sun*.[9] And there was finally another time, around the 1990s, in which some architects felt a renewed urge to conceive autarchic housing or deploy recycling construction techniques, which would potentially meet the needs of a new, widespread green sensibility.

9. See Daniel A. Barber, *A House in the Sun: Modern Architecture and Solar Energy in the Cold War*, Oxford University Press, Oxford: 2016.

10. See Jenna McKnight. "ZeroCabin in Chile designed to operate off the grid" Accessed 15 April 2020. https://www.dezeen.com/2020/02/29/zerocabin-chile-off-the-grid/

What was the common thread to all of these precursors of the present urgency? They were too early, they failed bluntly, and they were erased by history — at least until recent rescue was prompted by a new environmental agenda.

Caught in mourning for the demise of the architectural avant-garde, it saddened me that such attractive stuff for new manifestoes was now being left to thoughtful amateurs. It puzzled me that such an environment-prone component of architectural myth such as bold autonomous dwellings were now left to the musings of non-architects.

Today, the architectural avant-gardes may indeed be dead and buried, but there is a well-established architectural profession. When you read a description of a self-sufficient, low impact cabin that can "be tailored to the local climate and topography, along with a client's specific needs,"[10] you would expect architects to be around — at least to provide that familiar lingo. Even if the little cabin did not scream of revolutionary forms, you would suspect a few architects to be involved in the mix.

Yet, as underlined in architectural outlets, the ZeroCabin project was created by a multidisciplinary team of six people,

Global Earthship by Michael Reynolds of Earthship Biotecture

none of whom had "formal training in architecture." Rather, they had backgrounds in civil engineering, physics, biology, chemistry, sustainable farming and, eventually, interior design.

I remained perplexed at this utter absence of input from the architecture realm, and had to dig a little deeper to finally discover some glimpse of an architectural input to the cabin project. As it turned out, one of the team leaders confessed his inspiration came from the hero-like protagonist of the 2007 *Garbage Warrior* documentary, American architect Mike Reynolds. This motivational figure, who is still building 'earthship' passive solar houses in New Mexico, ended up surfacing as the architectural key for the whole cabin endeavour. But he also emerged as a typical remnant of a lost environmental avant-garde, a momentous trend which now seems to belong in an alternative architectural universe.

Shortly after he graduated from the University of Cincinnati, in 1969, Reynolds had his thesis published by the Architectural Record and surfed the wave of a peculiar architectural avant-garde concerned with emerging environmental concerns. He was part of a wider movement that, during the early 1970s, responded to ecological impacts made clear by Rachel Carson's seminal book *Silent Spring*, in 1962, or, later, Buckminster Fuller's 1969 *Spaceship Earth*. He integrated a group of architectural visionaries that anticipated and acted on the implications of Club of Rome's report on the limits of growth, in 1972, or the major oil crisis of 1973. Long before the University of Auburn's Rural Studio made it a tenet of its design-build pedagogy, he deployed waste materials and soil to achieve what he still calls today "earthship biotecture."

As stated on a Wikipedia page, "with the rise in concern over global warming, Reynolds has been elevated as a prophet of the green movement." Indeed, with his direct recycling practices, Reynolds was an early advocate of ecological reduction and of the proactive reuse of common trash, such as aluminium cans and tires. With these, he made way for architectural insulation and thermal mass with soil filling, which he even patented for the United States. One could picture him as the inventor of the yet inexistent practice of architectural compost.

More importantly, as others at the time, he investigated how to make his structures off-the-grid, by deploying initial versions of solar panels, geothermal cooling or regenerative food production. The prominence of his work at a given period even prompted some Californian celebrities to commission him new projects. Given the experimental, anti-building code nature of his work, however, he ended up facing lawsuits, losing his architecture license, and heading into a depression. He was suffering the same fate as the 1970s environmental avant-garde: a fatal inadequacy to find its proper place within both the architectural orthodoxy and a consumption-driven society.

It definitely felt to me like Mike Reynolds' trajectory could provide some clues to that ultimate question of what ever had happened to the environmental avant-garde. And over a phone conversation, some clues did emerge.

As he confided, while a new ecological conscience was discreetly popping up around California, he was down in the deserts of New Mexico, much by himself, making his recycling experiments with what most would call garbage. Out of curiosity and a desire for an alternative way of life, others came and went. They would usually give up, as they got frustrated at their own inability to sustain themselves off-the grid. Reynolds, however, stubbornly stuck to his ideas, attempting to devise this architectural system that he would call "biotecture."

As Reynolds put it, at the time "they didn't realise it was going to take 50 years to figure out how to make buildings heat and cool themselves, how to make them catch their own water, how to make them treat their own sewage, how to make them make electricity, how to make them produce food, how to deal with this natural resource we call garbage."

Thinking ahead of technological possibilities can make it hard to realise your visionary dreams, as logical as they may sound. If you are only part of a cultural avant-garde, you may expect ideas to produce immediate effects as they are reproduced by others. However, when your forward thinking implies immediate practical achievements, your commitment to a certain way of thinking may inevitably fizz out before you reach your goals.

And yet, as Reynolds said, those precursors "served a purpose." Even if experiments like his did not always work, indeed "their ideas, their thoughts, their reactions were relevant." It was worthwhile to persist.

It was only much later that he found that, with changing conditions, more people were "lending an hear" to that anticipatory way of thinking. As he noted, "we're ruining the place where we are living, and it is showing." Through the media, people are gaining a new awareness about ecological issues. "So," he concluded, "people are again turning to these directions, not from a moral standpoint, but because they feel insecure about the way things are."

At the time of his early work, however, Reynolds was being ridiculed and condemned. "I was called a disgrace to the architecture community because I was building out of garbage," as he recalls it. But, while working isolated, he had enough praise and recognition coming his way to make him persist. So, even if he wasn't aware of the avant-garde spirit of his practice, he just kept on doing it.

Unknowingly, Reynolds was part of an alternative, transient architectural movement that a recent exhibition dubbed "Hippie Modernism." Protagonists of this 'drop-out' version of modernism, as Andrew Blauvelt calls it in the exhibition catalogue,[11] included artist-architects such as Ant Farm or Site, Vienna-based Haus Rucker-Co, Florence-based Superstudio, Californian architects Sim van der Ryn and Lloyd Khan, but also 'zomes' inventor Steve Baer or Cal-Earth Institute founder Nader Khalili, instigator of another earth-based architecture practice still active today.

11. See Andrew Blauvelt (ed.), *Hippie Modernism: The Struggle For Utopia.* Edited by Andrew Blauvelt. Walker Art Center, Minneapolis: 2015.

These pursuers of alternative architectural practices were not only proto-environmentalists, but part of a wider counterculture that, out of the social conflicts of the late 1960s, and from different fields, questioned organisational models of post-war society. So they fought the political and artistic establishment, as well as what was seen as a destructive consumeristic society.

As Blauvelt notes, this counterculture, like others in the past, operated through "a diverse range of refusals." In the architectural realm, their refusals included the rejection of 1960s High Modernism, with its "lack of social engagement," as well as its growing distance from any holistic vision of nature and technology.

Only informally did they aggregate as a neo-avant-garde. They came up with a new aesthetic and radical practices that, after the period's *zeitgeist*, sought to catalyse "a new social, cultural, political, and ecological utopia." As they "challenged the disciplinary boundaries of architecture," they also opposed those "late stages of a modernism that had failed in its earlier utopian, avant-garde promise of social transformation."

In California's Bay Area, these avant-garde architecture environmentalists arose with a strong emphasis on ecological issues. In Europe, on the other hand, ecological concerns were

Vienna-based Haus Rucker-Co
Oase No. 7 (1972), Created for
Documenta 5 in Kassel, Germany

but a part of a larger agenda, which included the contestation of architecture's static monumentalism through performative and perception-altering actions. Haus-Rucker-Co's intervention at the referential 5th edition of Documenta, in Kassel, Germany, consisted of a bubble 'environment' that not only addressed issues of pollution and self-sufficient ecologies, but also commented on the institutional character of the building to which it attached itself as a parasite.[12] As for Superstudio, they went as far as refusing design and formal structures as expressions of consumer culture — leading them to imagine dystopian ecological scenarios as a form of social and political critique. In common, however, all these architecture collectives shared a nascent awareness of environmental catastrophe, and a need to respond to it with radical new practices.[13]

In this sense, another contributor to the *Hippie Modernism* catalogue points out that this counterculture's "design efforts in ecological sustainability and classroom liberation invite comparison to the reformist programs of previous modern movements."[14] As Greg Castillo goes on to suggest, similarities of this movement's pathos to the Bauhaus' joyful Gesamtkunstwerk are defensible. But only if underscoring that interdisciplinary collaborations were directed towards holistic and critical considerations of nature and Earth's wholeness, rather than just embarking on a design-based wish for "aesthetic totality."

At the core of Castillo's historical account are the counterculture battles that opposed these early environmental innovators to other critical thinkers of Western society at the time. The events at the Freestone Conference, a seminal gathering that took place north of San Francisco, in March 1970, reveal that the specific concerns of the environmental neo-avant-garde were much ahead of their time, even when it came to other forward-looking political positions.

While the hippie designers "embraced recycling and bricolage as totems of a post-Fordist culture," they invited the attacks by intellectuals that remained referential in the academic

12. For further insights on the environmental strategies of Haus-Rucker-Co, see Esther Choi, "Atmospheres of Institutional Critique: Haus-Rucker-Co's Pneumatic Temporality," in *Hippie Modernism: The Struggle For Utopia.* Edited by Andrew Blauvelt. Walker Art Center, Minneapolis: 2015. Also available at estherchoi.net, Accessed 16 May 2020, http://estherchoi.net/index/atmospheres-of-institutional-critique-haus-rucker-co-s-pneumatic-temporality

13. For an interesting analysis of Superstudio's work as avant-garde, see Ross K. Effline, "A Tentative Embrace: Superstudio's New Media Nomads," in *Regarding the Popular: Modernism, the Avant-Garde and High and Low Culture*, edited by Sascha Bru, Laurence Nuijs, Benedikt Hjartarson, Peter Nicholls, Tania Ørum, Hubert Berg. Walter de Gruyter, Berlin: 2011.

14. See Greg Castillo, "Counterculture Terroirs, California's Hippie Enterprise Zone." in *Hippie Modernism: The Struggle For Utopia.* Edited by Andrew Blauvelt. Walker Art Center, Minneapolis: 2015. p.90.

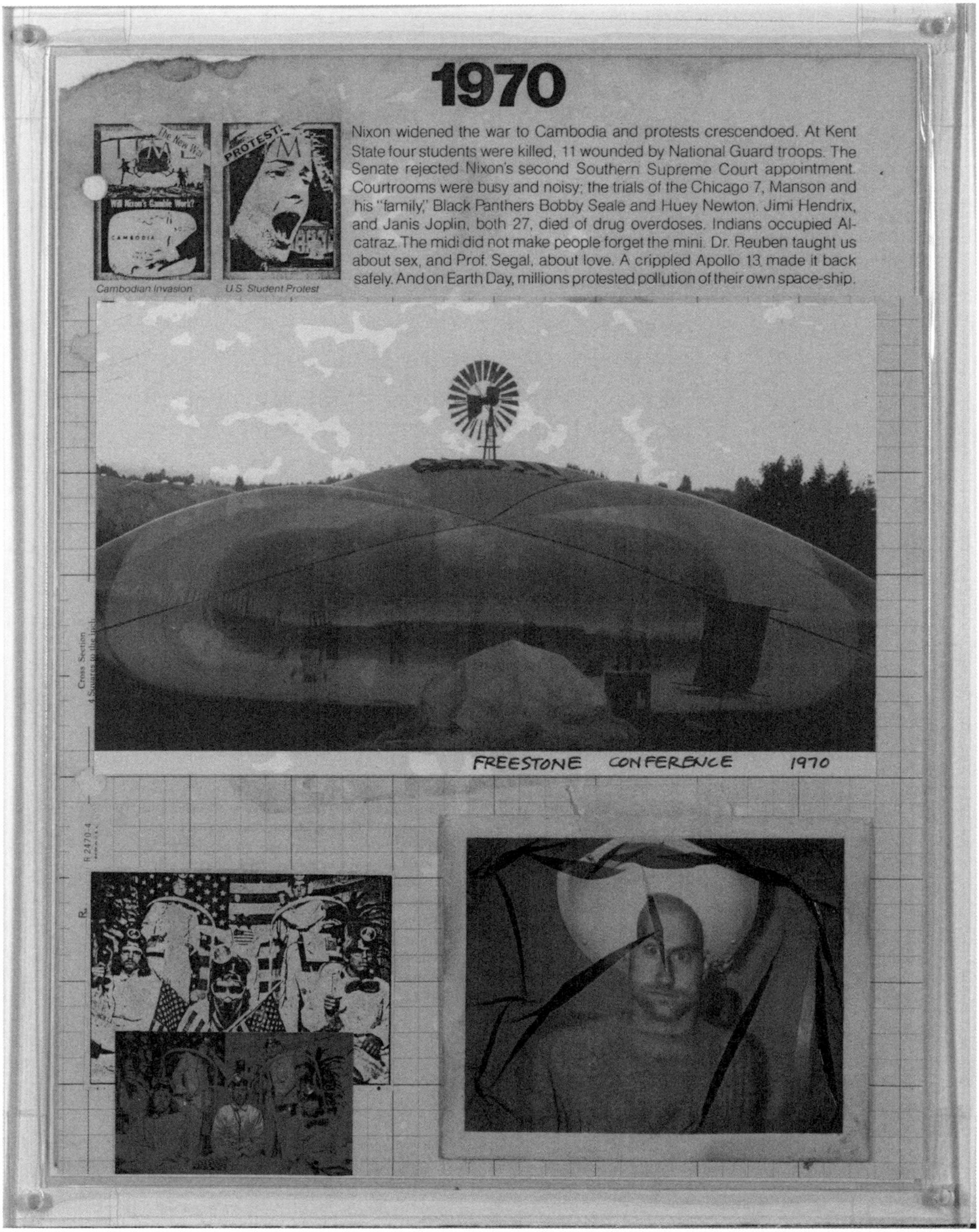

Freestone Conference 1970
(Ant Farm Timeline)

establishment of the coming decades. As Castillo reports, while Reyner Banham showed his "disdain for ecofreaks that were 'getting preachy'," Peter Hall demanded that "they stop talking about the coming apocalypse, alleging that the Bay Area environmentalists were simply hypocrites." And if that sounds incredibly akin to contemporary climate denialism, the final blow was yet to be perpetuated by French sociologist Jean Baudrillard. The New Left representative defended his own understanding of the political role of the avant-garde by fiercely attacking ecological activism. He claimed its environmental strategies, such as addressing pollution or announcing an ecological catastrophe, "were pure social manipulation" and "a new opium for the people."

In hindsight, one recognises Baudrillard's argumentation as the expression of a time before extractive practices and consumerism went beyond the French intellectual's wildest dreams. But it also revealed how foreign and esoteric those ecological concerns were half a century ago. This was the expression of a truly visionary perception of potential ecological disarray, which still today is barely catching up with mainstream discourse.

As Castillo notes, Baudrillard, Banham and Hall were all proven mistaken, and "the ecological crisis was neither a hippie hallucination nor an illusion produced by Marxist false consciousness." However, with its blatant failure to cater to the support of critical minds of the period, it was no wonder that, despite its clairvoyance, this environmental avant-garde was soon to fall into oblivion. With its failure to get traction even if only within architectural discourse, this alternative form of practice was clearly condemned to fail the battle with the self-perpetuation of mainstream trends of architectural modernism.

As we will revisit in a later chapter, late 1970s post-modernism was but a fleeting aesthetic bump in the road of modernism's way forward. Similarly, the environmentalist avant-garde was also a brief and marginal deviation in that heroic historical course.

Apparently, such freak deviation would leave no impact in mainstream architectural discourse, had history not shown that its themes were not to go away easily. Even if, like capitalism, modernist architecture proved highly effective in eliminating alternative models of practice and becoming utterly global, the seeds of this eco-modernism lay buried, waiting for a wake-up call.

Breeding on modernism's apparent success at formal self-preservation, a steady discourse on architecture's aspirations for autonomy has also thrived to this day. And that has essentially guaranteed that any calls for practice to again find its intrinsic motivations in everyday urgencies — as it is typical in avant-garde thinking — were hygienically kept at bay. As these orthodoxies went, essential drives for architectural creation had to be found within its own formal and typological history, within its aesthetic traditions and within the repetition of accepted formulas. Thus, any alternative interpretation or critique of a consensual notion of modernity was suppressed. While seen as steered by external forces, environmental avant-gardes too were to be buried, thus leaving architects unprepared to accept ecological concerns as potent creative drivers for architectural practice.

As the climate emergency became more evident, however, the motivations and prompts behind that specific strain of the architectural neo-avant-garde were back in full force. And so our attention was driven back to those early ecological visionaries.

As Castillo concludes in his essay, "whatever might be said about their motley raiment, grab-bag philosophies and clown-car comportment, ecofreaks reassessed as hippie moderns demand re-evaluation as an avant-garde galvanized by the radical mission of inventing environmentalism's everyday material culture." Further, this environmental avant-garde also encourages a re-appreciation of how design practices can be at the forefront of climate action, both in terms of activism and the integration of technological innovation in ecologically driven projects.

Seen from today's perspective, it is only understandable that there is again an appeal to counterculture movements of the early 1970s.

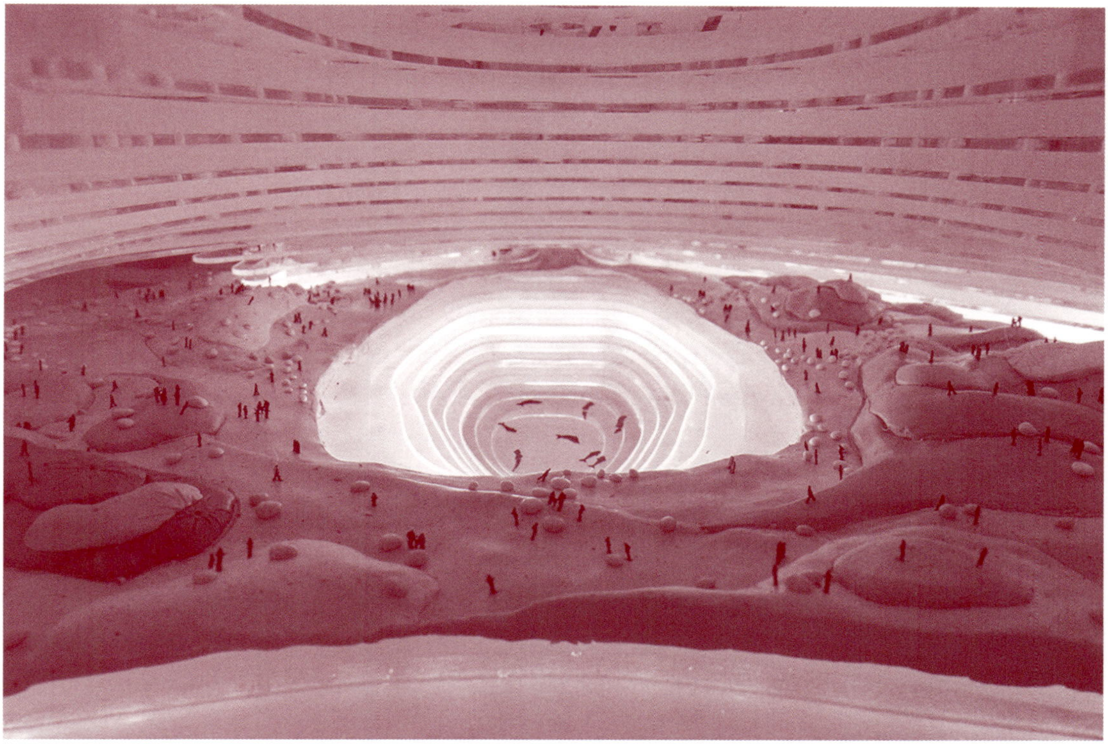

3.C. City: Climate, Convention, Cruise
© WORKac.
Photo by Bruce Damonte

Faced with the global impacts of an increasingly savage system of consumption, we can again feel attracted to move forward by taking a step back, looking instead for exploratory experiments in self-sufficiency. Faced with the mounting evidence of a pressing ecological distress, even as suddenly expressed in unexpected pandemic outbreaks, we can again feel the plea of refusing business-as-usual in any sector of practice, including the creative and design ones. In the midst of the successive crises of advanced capitalism, we may again lust for the bold creative ingenuity of the avant-gardes. We may indeed wonder where that avant-garde spirit has gone, and ask ourselves if we may hack some of it back.

Now, perhaps the messy, hippie looks of the early 70s avant-garde do not appeal to our sophisticated, stark aesthetic sensibility. But, as we now know, remixes are never quite like the original.

Think of pop music re-edits of vintage 1970s funk, and you may quickly reckon that what they lack in dirty authenticity, they conquer in subtle adaptation to contemporary taste. We will never go back to "hippie modernism," but we may well enter into a fruitful conversation with it.

One good example of such dialogue appeared during the first Chicago Architecture Biennale, in 2015, taking the form of a collaboration between New York practice WORK AC and the remaining elements of radical practice Ant Farm, Chip Lord and Curtis Schreier.

While Ant Farm dispersed in 1978, after their archive burned down, the interest of WORK AC in their countercultural explorations prompted the recreation and upgrade of some of their lost projects. Thus appeared *3.C. City: Climate, Convention, Cruise* — a new design intended to bring earlier ideas up to date in a context in which "climate change is challenging architecture to engage in urgent diplomacy and become an agent of change."[15]

15. See Amale Andraos and Dan Wood, 3.C.City. Work-AC. Assessed 16 May 2020. http://work.ac/work/3c-city/

Described as a hack of Ant Farm's polemical proposals at several scales, the new floating city concept most notably evolves to "facilitate dialogue and debate between people and other species." While Ant Farm's avant-garde inflatable aesthetics guides the design of this "vehicle of dreams," the current Anthropocene discourse on interspecies conviviality discretely informs the conceptual remix. As noted on WORK AC´s website, "solar panel shingles, pockets of greenhouses and gardens, an algae farm for biofuel and a water-collection river" are added to provide for "a new symbiosis between ecology and infrastructure." Importantly, green infrastructure is newly welcomed as a structural part of architecture's basic vocabulary for form-making. The dialogue with the representatives of the previous avant-garde has indeed produced a breakthrough.

So, we should celebrate these new "vehicles of dreams," which allow us to revisit visionary ideas of the past. Yet, it is still also relevant to ask again where the environmental avant-garde ended. Not because of some nostalgic notion of the role and

Convention City 1976
by Ant Farm

potential of the avant-gardes themselves, but because the early environmental architecture responses to the ecological crisis are still mostly valid today. Hippie mystique aside, the philosophical principles that regard the holistic integration of ecological concerns into architectural design are still sound enough.
So, we can start by revisiting their failures so as to simply guarantee that, this time, we get it right.

We can indeed recover some of the 1970s counterculture early environmental insights, as well as their political prowess, and update them to inform our current technological and design capacities. Even as we have now gathered how to deconstruct its defaults and failures, we can still learn from the lessons of the environmental avant-garde. Like the interdisciplinary team that came up with the ZeroCabin, today's architects can also find inspiration in the stubborn hopes and critical thrust of our ecologically concerned forebears.

Then, it is only a question of asking how we actually start.

2.
Start the Change

Climate Activists Protest Against Senate Leader Chuck Schumer Demanding He Sign On To Green New Deal by Drew Angerer

The early 2020s. Architecture discourse has decidedly moved online. In widely read architecture newsletters, the topic of climate crisis has also definitely jumped to the headlines. An expectable opening statement for an article would read like this: "Discussions regarding climate change and the need to make progressive reforms in the built environment are everywhere these days." Repeated exhaustively, this idea matched a peak of global awareness on the ongoing climate emergency, which architectural discourse could no longer ignore.

See Katherine Guimapang, "Kilograph and Climate Cents Team Up To Rewrite L.A.'s Climate Action Narrative." Archinect, Accessed 6 May 2020. https://archinect.com/features/article/150176134/kilograph-and-climate-cents-team-up-to-rewrite-l-a-s-climate-action-narrative?

No more than a couple of years. That was the time it took for a twist in the debate. Within those two years, Greta Thunberg, the Extinction Rebellion and the Sunrise Movement had emerged as activists with added sophistication and media impact. It became increasingly harder for architects to kick the can of the climate crisis down the road. Concerned scientists finally came out of the fear-spreading closet. Mainstream media embraced every angle of an ongoing environmental calamity. It was becoming definitely trying for the design fields to ignore the issue.

As politicians took due notice and started rolling out one Green New Deal after the other, it was proving steadily taxing for the architecture conversation to cling to a battered, corporately hindered, LEED-oriented notion of 'sustainability.'

Voices everywhere claimed we could no longer wait. A difficult international agreement had been reached in Paris, committing 197 countries to keep global warming under 2°C above 1850 average temperatures. Reducing greenhouse gases became an imperative, with CO2 emissions as the prime target. Addressing the climate crisis became an urgency. To reach goals set for 2030 and 2050, the change had to start immediately. Inaction was no longer tenable.

So, finally, the writer of that online newsletter newly coupled 'sustainability' with the wishful idea of 'climate action.' These were the two talking bullets that were now "difficult to avoid."

Yet, in the architecture field at large there was still little more than an inroad into a 'conversation,' and one that hid less comfortable facts. Architectural discourse was at a point at which it was becoming increasing difficult to defer the admission that, other than sustainable, built architecture was actually an active contributor to the climate crisis. As we knew it, 'sustainability' was no longer a sustainable discourse. Things had just gotten more complicated.

It was surely positive that the discussion had suddenly become inevitable. 'Progressive reforms' and 'climate actions,' whatever still unknown forms these might take, were now inescapable.

The new hot question, alas, was why is it so difficult to start addressing inevitable change in architectural practices.

To be clear, the scale of change required to turn built environment practices on their head practically overnight was and is obviously overwhelming. Around the time of that article, the various Green New Deals proposed by the European Commission and Democratic presidential hopefuls in the United States advanced goals that scientists deemed unfeasible.

At the table were plans to fully decarbonize Western societies within a period of ten to twenty years. A horizon of two decades to also decarbonize a sector, that of the construction industry, which had enduringly contributed around 40% of global carbon dioxide emissions.

That sudden decarbonization was not achievable through 'progressive reforms.' That meant an overhauling revolution in a sector that, much like oil, was highly conservative, generally driven by unadventurous market forces, and also supported by very strong lobbies. Surely, the construction industry was not going to let go of cement or steel over the next couple of years. And this would indisputably limit the scope of change or progressive reform that could be introduced in a professional field such as that of architecture. Without a general overhaul of behaviors of consumption and demand, architecture practice would have a hard time achieving a necessary transition toward a 'net-zero' footprint.

Although some of its academic sectors still strive to believe that architecture can be produced in a vacuum, when it comes to actually leaving a mark on the built environment, we know that this is not true. Architecture is built in an intimate dependency with the construction sector. It relies on materials and techniques that are provided by a market economy. It dangles on demand for new buildings, new urban areas.

And, even as a professional practice, architecture and its capacity for change will be bounded by its own set of restraints. It will be deterred by entrenched habits, old-school ways of thinking, and resistance to cultural adjustment. But it will also be inhibited to adopt drastic changes because of the simple need to sustain its intrinsic economic survival.

I can tell you a personal story that perhaps illustrates how difficult it is for architecture to start the change that was made inevitable by the climate crisis.

Back in 2017, I engaged in extensive research and readings for a curatorial project called *Eco-Visionaries*. At that time, I too became aware of the scope of change we required, if indeed we were to tackle a deeply scary future scenario.

With information overload sometimes comes painful enlightenment. As many others around this period, I became more vocal about climate change. I became aware that this was *the* issue to which I had to dedicate all of my capabilities. I inevitably made it the unfortunate topic of many of my social interactions — and of *all* of my dinner conversations. I also made it central to any new professional endeavour I would embark on.

At some point, an architect from the Boston area heard my usual rants. He was notably impressed with the passion I put into my description of how the architecture field needed a radical fix. He promptly invited me to come and do a presentation to his office. And, then, I was paralyzed.

I was not paralyzed because I couldn't eloquently describe all the reasons why architecture was an inadvertent accomplice of ecocide. And if I had doubts how impacting a case I could make, I could always resort to handing out copies of David Wallace-Wells' *The Uninhabitable Earth.*

By then, I could certainly make a very good case of why we architects needed to urgently start changing our practices — at least if we thought that avoiding a cataclysmic environmental crisis could do with a little help from our profession.

No, I was paralyzed by the invitation for another, less obvious reason.

I was paralyzed on a moral issue. I was paralyzed because it seemed unfair to unload this huge burden of a responsibility on a bunch of architects who were just performing their business as usual, as best as they could.

I was perfectly able to deliver this urgent spiel to an unknowing audience. I had already realized that the bad news, as all the other bad news in the media, would soon sink into the oblivion of their everyday dwellings. But it felt harder to stand in a nice architecture office, take the higher moral ground, point my finger and tell these architects their practice was verging on a criminal attack on the planet's environment.

It felt like too much.

Who was I to preach radical change to architects who were diligently responding to the demands of their clients, their cities, their niche markets? Who was I to call for immediate climate action from architects who were as socially committed as they could possibly be in shaping the urban environment, in responding to the housing crisis or in infusing some remnant of spatial quality into fiercely commercial projects?

For sure, even if at a small scale, they could infuse their everyday decisions with incremental change towards more eco-friendly practices. They could engage in a choice of more sustainable materials. They could attempt to include a green roof. They could go on a bid for the client to accommodate the latest energy-saving technology. And yet, there would be structural constraints preventing them from fully starting the change that was urgently needed.

These structural constraints are perhaps what we first must disentangle and make clear, particularly if we want to see that change unfurl at the scale and speed that is demanded by a state of emergency.

Even with such constraints it was also true that, in this same period, there were already architectural offices starting to commit to design only zero-emissions and even negative-emissions buildings — as I will come back to in more detail. However, these were either star-architects or activists, not exactly your average, middle-of-the-road architectural office.

As any follower of French sociologist Pierre Bourdieu could explain, global architects departed from positions in the architectural field that allowed them to pioneer change. They could even make that change an asset and a hallmark of their recognition. The symbolic capital they had already accumulated could be safely invested on the right side of history, with the added benefit of underscoring their green leadership for the field at large.

In the case of star-architects such as Snohetta or Foster + Partners, the economic breadth of their global commissions made experimentation and risk tolerable. Their strong partnerships with top international engineering firms led them to naturally lodge the latest technological advancements in materials research or energy-saving strategies. They could guarantee an economic return from wealthy clients, who progressively welcomed the marketing mark-ups of adopting greener strategies in their next developments. Such pioneering clients could also trigger following at a wider scale.

Mitigation of Shock by Superflux

If cost-benefit analysis would prove fitting, less wealthier investors would also eventually be inclined to adopt similar strategies. Yet, while star-architects benefited from an upper-market logic, it was not obvious that such a logic could quickly trickle down to all levels of demand and, consequently, to the average architectural commission and your typical architecture firm.

As for activist practices, while they provide us with alluring examples, they remain even more of a niche activity, often exclusively supported by extra-market forces such as academic research, philanthropic foundations or circuits of cultural consumption. As many others, practices such as Andrés Jaque's Office for Political Innovation or London-based collective Superflux have become very agile in navigating biennales, triennales and other cultural venues for the presentation of innovative architectural ideas. Through such vehicles, they have created room to explore experimental concepts related to coming environmental threats. They too lead by example, testing the ground for future ideas, often inspired by out-of-the-box research and unusual topics. Unfortunately, not everybody will be able to draw on the limited economic resources on which these academic and cultural practices depend. But such activist forerunners surely contribute to the circulation of new scenarios and strategies in the architectural discourse.

Other practices strive to maintain a balance between a discreet activist streak and a solid economic position within a market setting. These are the practices that usually appear as committed champions of more mainstream notions of sustainability. In the United States, if you are a client looking for prospects of a greener building, you can now find those architects through the typical online rankings of "the most environmentally friendly architecture firms"[16] or "the top 50 firms in sustainability."[17] These are not necessarily star architects, but large or middle-sized offices that decided to distinguish themselves from competitors by their ability to engage new social demands with emerging engineering technologies. As a sense of responsibility towards climate action accrues in society, clients show new concerns. Be it wealthier individuals, corporations or developers, they soon aspire to have a range of architectural answers readily available. As the market grows, these well-positioned firms enjoy positive feedback loops by further enhancing their response to climate related challenges.

In parallel, there always have been charming calls for activist architecture out there. They are meant to tackle that rising social awareness of a given topic, but also to find to ways to stimulate it. They remind us that "the power of art as activism lies in the unique ability of art to break the emotional barriers of the observer."[18] They remind us that, as people want to show they care, they need a stimulus from creative practices to envisage how they too can contribute in lasting ways.

In a context of environmental crisis, everybody's efforts are called for. As that realization emerges, many of the people who want to contribute will also be looking to leave a visible mark. They will be eager to enjoy a symbolic social return from exercising their individual responsibilities. That translates into the patronage of innovation in addressing increasingly palpable problems.

16. See Tammi Edwards. "25 of the Most Environmentally Friendly Architecture Firms,' 26 Aug 2019. The Art Career Project, Accessed 25 Feb 2020. <https://www.theartcareerproject.com/blog/25-best-environmentally-friendly-architecture-firms/>

17. See David Hill, "Architect 50: Top 50 Firms in Sustainability" 08 Nov 2018. Architect Magazine. Accessed 25 Feb 2020. <https://www.architectmagazine.com/practice/architect-50-top-50-firms-in-sustainability_o>

18. See Niall Patrick Walsh. "We Need DIY Activist Architecture to Fight Climate Change" 24 Sep 2019. ArchDaily. Accessed 25 Feb 2020. <https://www.archdaily.com/925260/we-need-diy-activist-architecture-to-fight-climate-change/>

ISSN 0719-8884

This, in turn, opens up the way for smaller, more activist and experimental architectural offices to expand towards building commissions that effectively start to demand alternative solutions to decarbonization, ecosystemic balances and more of a positive, clean impact on the built environment.

In all these cases, architects set up models, demonstrate the feasibility of new techniques and pave the way for future economic adoption of otherwise niche solutions. They provide a fundamental background for ideas to replicate and eventually spread across the wider professional landscape. This is certainly one of the necessary pathways for effective change in a field of practice.

Yet, in order to achieve a massive overhaul of architecture practice as we have known it in the 20th century, it will take much more than to follow the example of a few innovators. While activism may provide glimpses of inspiration for many a practicing architect — and their clients — it is not so clear that it can fast become the tenet of mainstream architecture culture.

As any theory of change will make clear, it will take much more than having a consistent mass of early adopters. However, the speed of change that is required to simply address immediate needs of climate change mitigation is overwhelming. The quantity of carbon dioxide we must remove from the atmosphere, the built environment, and our own practices in a very short period of time, is overwhelming.

This means it will take much more than a reasonable number of architecture practitioners following on the heels of some well-positioned colleagues. It will take a whole new understanding of the architectural field, both in terms of the conceptual framework of its form-making impulses and, most importantly, in terms of the economic superstructure in which the practice is inserted.

As much as one critic has recently noted, "market forces do not conspire deliberately against architectural beauty." Similarly, market forces do not conspire against architecture doing a better job at addressing the impacts of the climate crisis. However, as Marianela D'Aprile continues in her piece about how "winner-takes-all capitalism undermines the real value of architecture," a market economy in a capitalist regime does "prioritize profit over good design."[19] So, while "most architects don't sit in their offices complaining about what capitalism is making them do," they do take such conditions as unquestioned compromises of the cultural logic that determines their field.

19. See Marianela D'Aprile, "How Winner-Take-All Capitalism Undermines the Real Value of Architecture," 12 April 2017. commonedge.org. Accessed 25 Feb 2020. <https://commonedge.org/how-winner-take-all-capitalism-undermines-the-real-value-of-architecture/>

Yet, when architectural practice risks irrelevance due to the complete overhaul produced by a climate emergency, architects may as well question the unsupportive economic system under which they operate. Just like under the pressure of profit they still strive to produce beauty, even if this is not as valued, they might also rediscover a deeper social meaning — and find an economic justification — for a more environmentally responsible design. They may be delivering commodities, but they may successfully argue that more sustainable commodities do produce an economic return besides the immediate ecological benefits. And this will be even more significant as clients progressively attach a symbolic value — or the offsets of a guilty conscience — to environmental features.

To start changing architectural practice as we know it, we need to grasp other structural aspects. For instance, it is urgent to realize that the required level of change is akin to the conceptual revolution that marked the introduction of modernist architecture and the International Style at the beginning of the 20th century. Some voices in the architectural field have already started to grasp and discuss this evidence. But that is only one way of looking at the scale of change we are facing.

In fact, what is required equals that historical realization *plus* the need to articulate it with broader social and political transformation. That is, a paradigm shift of the discipline is also required to dynamically interact with societal transformations which alter the nature of the demand and pressure put on architecture practices by their current economic contexts.

As we will see in later chapters, the change required in architectural practice cannot but go hand in hand with the massive cultural change implied by what many call the *Great Transition*, and what political pundits started to address with several, catchy *Green New Deals*.

However, if we want to consider how such transition takes over the design disciplines, we must acknowledge other obstructions out there. There are more subtle inhibitions that stop us from triggering the changes in architectural practice that suddenly became urgent. These may be surprising. And yet, they may also be very revealing.

When positing the need to start changing architectural practices, one can accept that the market forces behind architectural production must undergo a transformation of demand. It is much more difficult to apprehend, however, why it is that one environment that is relatively protected from market forces is not already taking the lead in instigating this change in the architectural field. It is much more difficult to understand why the education sector is not yet wholly committed to preparing architecture students to the inevitable change in the way they will practice architecture.

Ultimately, it is baffling that not many major architectural schools have yet taken the lead in making this an urgent subject of experimentation and design thinking for future practitioners.

The article with which I started made this much evident: it was increasingly "difficult to avoid talking about sustainability and climate action when discussing architecture." Yet, by the beginning of the 2020s, most architecture schools around the globe were actually doing a pretty good job at avoiding the issue. Many of those who were supposedly preparing new generations to face the design problems of the coming decades, actually seemed to be the first to resist change.

When near-future design problems would seem to be increasingly permeated by the effects of the climate crisis, such resistance sounds unreasonable. With international accords, activist and media pressure, professional bodies declaring the urgency of decarbonization goals, one would expect avant-garde pedagogical sectors would be quick to seize the opportunity to provide the right answers.

And yet, perhaps this resistance to change is easier to explain than expected. Perhaps, this actually sheds some light on why an essential part of the disciplinary superstructure of architecture repels environmental-related change in the realm of design practice. It is difficult to embrace a problem that, quite radically, puts your practice in question.

When I was granted a Loeb Fellowship and landed at Harvard GSD in late 2019, I couldn't but help noticing this dissonance.

I had presented the climate crisis as my subject of research. I was aiming at benefiting from the wider Harvard ecosystem to deepen my understanding of the topic from the perspective of different fields of knowledge, from science to policy-making. As such, it was inevitable to register that, at the architecture school, I showed up as a little bit of an alien.

For sure, outside the main architecture core there were side courses on urban adaptation or resilience, or even the occasional option studio on planning for rising sea waters. There was even a novel course on Climate Justice, in which I readily enrolled. To be fair, while I attended that academic year, from one semester to the next I could see many more of the option studios setting climate change as their theme of departure. This shift might have been driven by a strong presence of urban planning and landscape architecture in the GSD curricula. But it seemed to announce a turning point. Even if only by popular demand from the students, this shift surely reflected that inevitable and insidious turn of the general discussion towards climate issues.

Still, let us face it.

The core of the professorship that was engaged with ivy-league notions of architectural pedagogy was still circumventing the topic. As I sensed it, it seemed quite alright to have environmental concerns as yet another technical, extra-disciplinary contribution to the design process. Even if nonchalantly, ecological preoccupations from the students were accommodated as a sort of passing trend. Climate change seemed like one of those additions to ongoing discourse that one cannot publicly reject, but accepts as a slightly annoying diversion from more noble matters.

Ultimately, it felt as if climate change would endanger decades of careful construction of some sense of disciplinary resilience. It felt as if serious consideration of those ecological issues in the essentials of form-making would prevent the system of architecture from going back to some original, autonomous state of grace.

In other terms, one could argue that it is at the level of architectural education that other hidden truths come to the surface. These are the truths that come from the core of the discipline, and its self-assertions as a field of historical knowledge production. These are the principles that regulate the evolution of the field, the unwritten gate-keeping norms that sanction or limit what is to be valued in the discourse of architecture. In other words, these are the truths by which academia guides itself.

From this point of view, similarly to what was happening in the political arena, the postponement of a supposed ecological crisis was perfectly understandable. As one comes to the realization that many of the cherished principles driving architectural conception may be proven aberrant by environmental concerns, it is naturally tempting to keep these at bay for as long as possible.

If your aesthetic philosophy of architecture includes stark concrete surfaces, beautiful expanses of glass, luxurious materials brought from across the globe, bold cantilevered volumes at the cost of expensive engineering feats, you will probably want to resist the paradigm shift implied by the climate crisis. It will seem improbable that you will get your kicks out of a potential trash aesthetic, practices of recycling or humble returns to vernacular lessons on how to deal with passive energy systems.

If you are an architectural aesthete, accommodating the consideration of climate responses in your pristine architectural composition may be simply irritating. Augmenting the solar harvesting capacities of your building may imply certain forms, which do not necessarily agree with your acquired tastes. Including green roofs or other water retention mechanisms may not feel coherent with your amazing volumetric concept. You may simply feel insecure on how to design with mass timber systems, rather than sticking with those gorgeous metal structure details that you treasure in Mies van der Rohe. You certainly do not wish for the prosaic apprehensions emerging from the climate crisis to command your design strategies.

If one trusts the adage introduced by activist author Naomi Klein that climate change indeed *changes everything*, this is all understandable. It is only natural that the architectural field's gatekeeping devices will be immediately activated.

In face of potential disruption, defence systems will automatically rise as if combating a virus. When the building bricks of architectural discourse risk de-assemblage, barriers of cool indifference must be subtly erected. With a system of architectural thinking carefully built upon a glorious modern history, any threat to that history's morale stand must be avoided. And as much as it is now hip to speak of the decolonization of knowledge, it is complicated to see your discourse colonized by such an external might as that of the climate crisis.

So, I would defend that, if indeed we want to start that necessary change in architectural practice, we must begin by addressing the *status quo* of architectural academia. And this means several different things.

It may indeed mean disruptions in leadership of academic centers. Alternatively, it may also mean that the relevant players in that restricted field feel positively challenged to adapt — especially if they want to maintain their power. They may be driven to accept that, given the crushing weight of this external transformation, they are better off embracing such change as a positive force. They may be persuaded to become leaders of such a change.

So, fortunately, pledges for leadership in climate-related architectural education have started to pop up. A very few architecture schools around the globe have already sent the signs.

Think of what happened in London. The capital of the United Kingdom was one of the first global cities where new forms

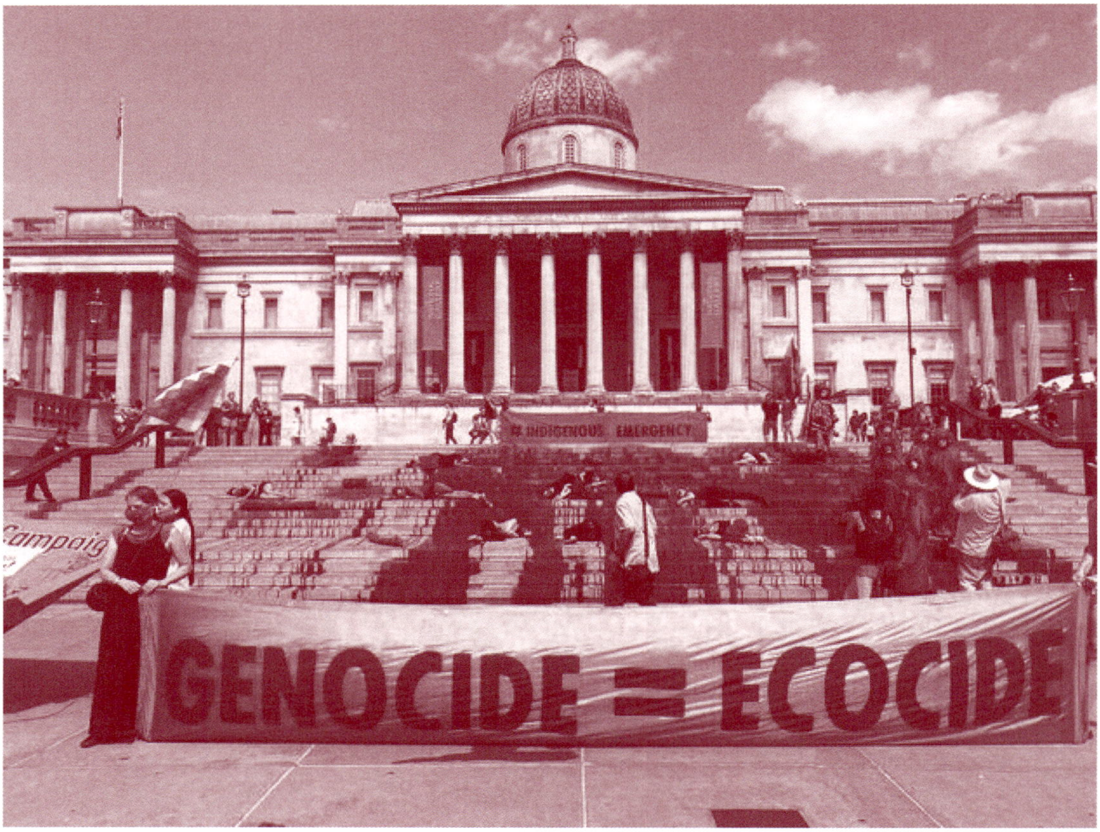

Extinction Rebellion
© Talia Woodin

of climate activism took shape, notably with the public appearance of the Extinction Rebellion in late 2018. With the visibility and impact of protests, institutions across the board quickly started to take due note of the transformations implied by the climate crisis.

On the 1st of May 2019, the Parliament of the United Kingdom was one of the first to declare an environment and climate emergency at the national level. Organizations and coalitions in the architecture arena soon followed.

That same month of May, seventeen winners of the Stirling Prize, such as David Chipperfield Architects, Foster + Partners, Rogers Stirk Harbour + Partners, Stanton Williams, WilkinsonEyre, or Zaha Hadid Architects, launched their own declaration of "climate and biodiversity emergency."

UK Architects Declare

They were bold enough to propose that "for everyone working in the construction industry, meeting the needs of our society without breaching the earth's ecological boundaries will demand a paradigm shift in our behaviour."[20] As such, among several measures, they advocated for "faster change in our industry towards regenerative design practices," but also "a higher Governmental funding priority to support this."

On 27 June, the Royal Institute of British Architects committed publicly to a climate and environment emergency declaration. This came complete with an immediate five-year action plan, including support for chartered practices to "work towards the whole-life net zero carbon standard." Along with the declaration, RIBA President Ben Derbyshire stressed how "we architects need to transform the way we practice and along with our fellow professionals around the world, make changes that will impact at a global level."[21]

In that context, it was not much of a surprise that, by October, one of the leading schools in the city, the Bartlett School of Architecture, also declared its "climate and ecological emergency." One of their first proposed actions was to establish "a working group of staff and students to explore how architectural education and research could be further developed and harnessed to avert catastrophic climate change and irreversible biodiversity loss."[22]

20. See "UK Architects Declare Climate and Biodiversity Emergency." Architects Declare, Accessed 28 Feb 2020 <https://www.architectsdeclare.com>

21. See "RIBA commits to climate emergency declaration and action plan." 27 June 2019. RIBA, Accessed 25 Feb 2020. <https://www.architecture.com/knowledge-and-resources/knowledge-landing-page/climate-emergency-declaration-and-action-plan>

22. See "The Bartlett School of Architecture declares a climate and ecological emergency." 4 Oct 2019. The Bartlett School of Architecture, Accessed 28 Feb 2020. <https://www.ucl.ac.uk/bartlett/architecture/news/2019/oct/bartlett-school-architecture-declares>

This coincided with the launch of another coalition — and its online petition-manifesto — in a summit called "Architectural Education and the Climate Crisis."[23] The organizers posited that "our education has a key role to play in addressing the ongoing ecological crisis," and came out with a "call for a curricula change." They further expressed their concern "that at present our education does not give sufficient weight to the inherently ecological and political basis of architecture, nor to our responsibility to meet our uncertain future with socially and environmentally informed practice."[24]
As I wrote, this petition counted with already nearly 2300 signatures from prominent architectural educators across the United Kingdom.

More interestingly, though, two years before this authentic cascade of climate declarations, another London school had already "sounded the alarm."

As soon as Adrian Lahoud settled into his position as the new dean of the Royal College of Arts, he started to make changes. In 2017, the school announced its new MA course on Environmental Architecture. As described, this was intended to "focus on the co-existence of different human societies with earth systems, landscapes, environments and ecosystems," beginning by "looking at the impact of climate change and its social and spatial consequences."[25]

Bringing with him a background of climate-related research, Lahoud's implementations underlined that effective responses to the ecological crisis in architectural education did require a committed leadership. Opportunities arising, individuals who were already engaged with the topic would be more prone to promote changes at the institutional level.

As suggested earlier, change often starts from the impulses triggered by activist pioneers who reunite the right conditions to set a viable model. Be it in education, media, professional organizations or cities, committed individuals have the power to start the change. And as declarations of emergency continued to spread, these models of climate leadership were bound to resonate elsewhere.

In the United States, New York City made its own move in June 2019. It became the world's largest city to declare a climate

23. See Architecture Education Declares, Accessed 28 Feb 2020. <https://www.architectureeducationdeclares.com/events>

24. See "Open Letter to the Architectural Community: A Call for Curriculum Change." Architecture Education Declares, Accessed 28 Feb 2020. <https://www.architectureeducationdeclares.com>

25. See "Royal College of Art launches masters degrees in environmental architecture and city design." 24 May 2017. Dezeen, Accessed 28 Feb 2020. <https://www.dezeen.com/2017/05/24/royal-college-art-rca-launches-new-masters-degrees-environmental-architecture-city-design-london-promotion/>

emergency. Many others followed. Professional organizations such as the Architectural League of New York, while not declaring an emergency, immediately started design-related cultural programs on the climate crisis.

By early August, when the next dean of the Tulane School of Architecture was announced, it would seem only logical that the change in leadership could focus on a clear embrace of climate action. Arriving in a place beleaguered by climate disasters, incoming dean Iñaki Alday was prompt to assert that the New Orleans school would focus "on urgent problems, not self-indulgent fictions."

Declaring his lack of interest in "the endogamic discourses that have occupied academia for decades," the new dean spoke of "the opportunity to define the role that architecture can take in facing climate change — including other ecological crises, as well as in the process of urbanization under these circumstances — and the challenges of social and environmental justice that follow."[26] Consistently, the landing page of the school's website now includes a Dean's Profile in which these issues are clearly stated.[27]

This is not to say that other education institutions were dormant.

Architecture schools with an environmental tradition, such as the architecture school at Penn University, had already started their soft change. They focused on curricula and allured staff that have been researching climate topics within the architectural field. They even extended their action to initiatives beyond the traditional academic scope.

By Autumn 2017, the Weitzmann School of Design was offering a graduate course on Environmental Building Design that stressed "the renewed urgency of environmental issues — from global climate change to resource shortages and 'net-zero' design."[28] The course description course certainly demonstrated a keen awareness of how "a wider range of skills is required to achieve real innovation and to meet the needs of clients in [a] rapidly changing field." Furthermore, in 2019, the Ian L. McHarg Center of Urbanism and Ecology was sharing online its *Project 2100: An Atlas for the New Green Deal.* Extending the educational reach, the project's

26. See Antonio Pacheco, "We urgently need to change the way in which we are inhabiting the planet," an interview with Tulane University's new dean, Iñaki Alday." 5 Aug 2019. Archinect, Accessed 28 Feb 2020. <https://archinect.com/features/article/150145507/we-urgently-need-to-change-the-way-in-which-we-are-inhabiting-the-planet-an-interview-with-tulane-university-s-new-dean-i-aki-alday>

27. See Tulane School of Architecture website. Accessed 28 Feb 2020. https://architecture.tulane.edu/about/dean-alday

28. See Weitzmann School of Design Graduate Architecture website, Accessed 9 Mar 2020. <https://www.design.upenn.edu/architecture/graduate/master-environmental-building-design-mebd>

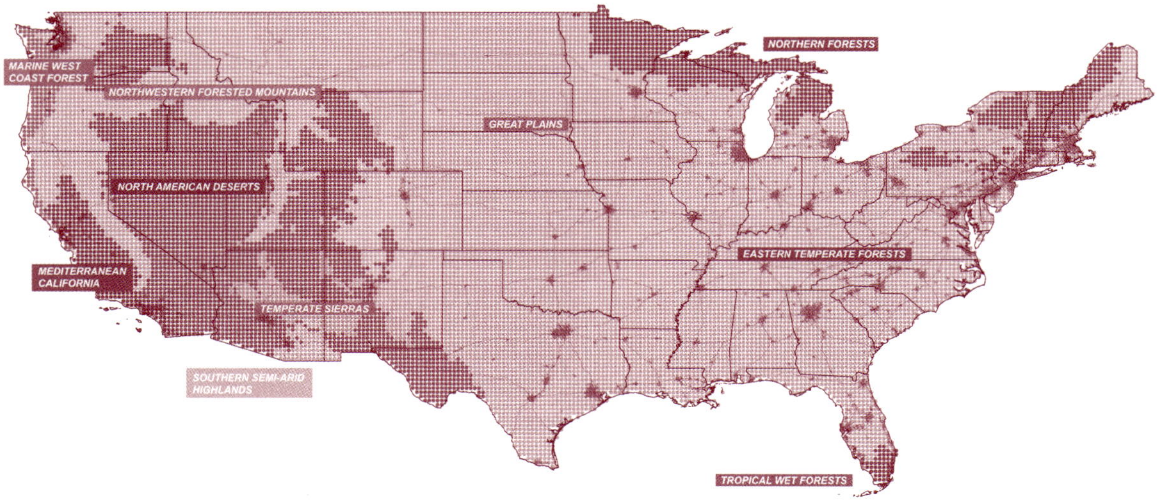

The McHarg Center and Billy Fleming, "An Atlas for the Green New Deal," 2019

maps on the effects of climate change were offered as "a resource for policymakers, planners, and communities."[29]

Along these lines, as from 2020 Columbia University's Temple Hoyne Buell Center for the Study of American Architecture partnered with the Association of Collegiate Schools of Architecture to start a prize for pedagogical innovation in the arena of "architecture, climate change and society." As it was publicised, the idea was to support and stimulate the creation of new courses on the topic "at a wide variety of architecture schools in the coming years."[30]

As stated in the prize announcement, it was understood that the focus on the urgencies of climate change was "antithetical to humanities-based critical inquiry" — if not to architectural discourse's preference for a solid disciplinary autonomy. Thus, the $10,000-a-piece yearly cash award went for five courses that directed at architecture students "the hard questions being asked of society as it grapples with the causes and effects of climate change." The selected innovative courses were described as seeking "the kind of realism that redefines problems and leaves room for the imagination." Topics ranged from the inevitable, hardly still avant-garde, "adaptation to sea level rise adaptation" to the more up-to-

29. See Emma Natanzon. "UPenn maps out climate change in support of a Green New Deal" 27 Dec 2019. The Architects Newspaper. <https://archpaper.com/2019/12/upenns-mcharg-center-climate-change-map/>

30. See "2020 Course Development Prize in Architecture, Climate Change, and Society," e-flux.com, Accessed 16 May 2020, https://www.e-flux.com/announcements/315374/2020-course-development-prize-in-architecture-climate-change-and-society/

date offerings on "public issues, climate Justice, and architecture" or "public architecture and the post-carbon imaginary."

Adding to such emerging trends, there are, of course, also the universities in which a strong tradition of environmental studies in different academic areas ended up determining new ambitions for their architecture departments.

This is the case of the Montreal-based McGill University. There, amongst a number of environmentally oriented architecture courses, a willingness to further strengthen its curricula and staff in that direction made it bring on board people such as Kiel Moe, who as Gerald Sheff Chair of Architecture now leads his department around the notion of "ecological construction." Moe was already a believer that there were flaws in architecture's typical understanding of energy and thermodynamics. He made it his research goal to better understand the economic, environmental, and ethical components of energy in the realm of building technologies — thus intendedly placing McGill's School of Architecture "at the vanguard of sustainability and activism."[31]

Similarly, halfway through 2019, UC Berkeley announced that one of its 1996 design graduates, author and practicing architect Vishaan Chakrabarti, was the new dean of its College of Environmental Design. As highlighted, the new dean would lead the school as the institution was now declared "poised to take on the three pressing challenges of our era — climate change, social inequity and technological dislocation — through the lens of design excellence." Never mind that Berkeley had once been the hotbed for California's environmental avant-garde.[32]

All these cases reveal that, if in embryonic form or following the examples of others, architecture schools will inevitably be waking up to the realities of the climate crisis and its impact on architecture practice.

And with that change, they address not only the educational needs of their students, but also the role of a more transversal cultural upbringing. They organize events, discussions, summits, even if these tend to remain in the academic circle.

31. See Peter Farbridge, "Construction Ecology: An Authentic Path to Sustainability," 30 Jan 2020. mcgill.ca, Accessed 16 May 2020, https://www.mcgill.ca/engineering/article/news/construction-ecology-authentic-path-sustainability

32. See Ryan Deffenbaugh, "City Urban Design Advocate Takes California post," 27 June 2019, Crain's New York Business. Accessed 16 May 2020, https://www.crainsnewyork.com/real-estate/city-urban-design-advocate-takes-california-post

However, information does percolate in other ways. After all, practicing architects will also be the educators of their communities and clients. They bear the responsibility to learn the lessons of environmental change for their practice. But they must also spread those lessons in their daily interactions with clients, with the construction sector, with society at large.

A sweeping change in academia's relation to the prominence of the climate crisis is important. And this is not only because schools are supposed to be training students for new conditions of architectural production. It is also fundamental because academic discourse eventually — and hopefully — trickles down to practice, providing guidance and research-based knowledge.

In reverse, architecture education has always drawn on practice. And this means that, in the midst of change that must come from strategic leadership at the top, schools can re-evaluate the balance of the teaching staff they bring on board. They can more actively cater for those practitioners that have been successful at allying form-making abilities with effective research into issues arising from the encircling environmental crisis.

Architecture schools should also be incorporating new theoretical approaches to the broader ecological crisis: the kind of urgent thinking that allow us to fathom, contextualize and integrate the different impacts of that crisis in relation to traditional guiding principles of architecture. In one sentence, architecture education should be taking the lead to start the change in facing up to the environmental crisis.

As such, it remains notorious that, as I was writing, no Ivy League school within the influential American context had yet announced a change in direction. And this is only relevant because other schools still look up to these as role models.

But this crisis will not magically wane by tomorrow. So, change may start elsewhere and leave some fretting behind.

3.
Weapons of Ecocide

Screengrab from Adrian Lahoud's video installation "Climate Crimes"

When one wandered through the colourful paraphernalia of *The Future Starts Here*, an architecture and design show at the Victoria and Albert Museum, one might just miss this particular exhibit. In the exhibition's climate section, there was a suspended geodesic dome in which dreamy graphics came in and out of focus. If you were lucky enough to look up at the right moment, you could have been mesmerised by images of mysterious gases evolving over the Earth's surface.

What was intended as a compelling immersive installation was somehow lost in the middle of so many objects and ideas vying for your limited attention span. Yet, you may still have been puzzled by those intriguing visualisations of our planet, and asked yourself what they wanted to tell you. As often happens in today's museum's shows, if you wanted to satisfy your curiosity, your best option would probably be to try and look for the object's caption for some clarification. This would hopefully make clear what that installation was trying to convey.

Climate Crimes, as the installation is titled, was conceived by architect Adrian Lahoud, who has been referred here as the dean that brought the climate change agenda to the Royal College of Arts, with artist Michaela French. As we can still read on the exhibition's website, the installation intended to explore "the complex relationship between air pollution and the migration of refugees."[33] Through the use of existing climate visualisations, it sought to illustrate "how atmospheric particles originating in the wealthy nations of the global north — Europe, USA, China, and others — impact the global south, contributing to desertification and migration."

What grabbed my interest in the project's description, however, was that it was inspired by a Sudanese diplomat's accusations of an ongoing 'climate genocide' in Africa. The contentious claims were presented by Lumumba Di-Aping against the G20 countries, during the 2009 United Nations Climate Conference. The event was one of the first public instances in which climate change was argued to be a potential basis for crimes against humanity. Following on this, the authors of *Climate Crimes* hint at the idea that the climate models they use for their immersive experience "are not only a medium of decision making about the future, but also a future archive of evidence for present crimes."

In between these two moments — between the bold political intervention of a decade ago and the recent artistic reflection — a path opened. Climate crimes became a central part of what has crystallised in the international arena as a pervading notion of ecocide. Defined as a continued criminal attack on natural environments, ecocide has been advanced as an equivalent to genocide, albeit directed by humans at other species or specific ecosystems. The late law activist Polly Higgins (1968-2019) was one of the most visible proponents of this new crime. Defining herself as an Earth Lawyer, she fought for the inclusion of ecocide in international legal frameworks as "the extensive damage to, destruction of or loss of ecosystem(s) of a given territory, whether by human agency or by other causes."[34]

33. See *Project focus | Adrian Lahoud: Climate Crimes*, Victoria and Albert Museum. Accessed 3 May 2020. https://www.vam.ac.uk/articles/project-focus-adrian-lahoud-climate-crimes

34. See Proposed Amendment to The Rome Statute. Ecocide Law. Accessed 5 May 2020, https://ecocidelaw.com/the-law/what-is-ecocide/

The generalised adoption of ecocide as a basis for legal action and criminal liability still remains in the realm of environmentalists' dream quests. Yet, when it does become a legal standard, as it surely will when the going gets tough we will be looking back at our recent history with a very different inquisitive eye.

A few thinkers have already played with this idea: one day, future generations of humans will look back to our epoch and will be appalled by the expanse of destruction driven by our obsession with progress and growth. They will be mourning the radical loss of biodiversity, the large extensions of dead soil and dead waters, a planet ravaged by extraction. They will be wailing the devastating daily impacts of climate change gone wild. And they will be looking to prosecute and condemn the historical culprits of such broad ecocide.

As much as we would wish this would be only science fiction, this scenario could actually be closer than we think. That notion of ecocide, or the deliberate destruction of the natural environment, is already here. And it is gaining traction and triggering action. And that may quickly change the way we look at many activities that, until now, we appreciated independently of their environmental impacts.

History is often revised due to shifting perspectives, and the magnitude of the coming ecological distress will certainly lead to crucial re-readings of ideas we once considered benign and progressive. As ecocide makes it into broader legal discussions, it will surely boost a wave of lawsuits and even criminal prosecutions that has already started against polluting and extracting industries. In fact, the ecocide eradicating campaign launched by Higgins precisely decided to establish a legal precedent by targeting the Dutch oil giant Shell. Teaming up with the Ivorian community, they took the company to court in London and, after years of struggle, they achieved a significant success in 2011 making Shell legally accountable for a massive oil leakage and the ensuing environmental damage in the Niger Delta.

As reported in a United Nations University newsletter, this did not "undo the decades of social and environmental

Shell's 2011 massive oil leakage in the Niger Delta

injustice caused to the Niger Delta's communities and natural environment." But it did provide a well-publicised stepping stone in achieving the criminalisation of ecocide as a preventive measure that, in the future, may "ensure company CEOs, heads of banks and heads of state will not conduct business, finance, or support activities that could result in ecocide."[35]

If one remembers when environmental movements started rising, it will be no surprise that the attempt to inscribe the notion of ecocide in international law dates back to the early 1970s. It has since been discussed as "the missing 5th crime against peace,"[36] along with genocide, war crimes, or crimes against humanity. Although supported by many proposing countries, it was left out of a 1996 UN resolution on Crimes against Peace and Security of Mankind, which later became the Rome Statute of the International Criminal Court. It was eventually recognised in the laws of ten countries around the world. The first of those was Vietnam, whose 1990 Penal code states in its Article 278 that "destroying the natural environment, whether committed in time of peace or war, constitutes a crime against humanity."

Given the existing legal framework, perpetrators of genocide usually face trial within years of their actions. In a context of ecological emergency, with law evolving according to shifting sensibilities, it could be sooner than we think that multiple enactors of ecocide will be brought to court. As mentioned above, activists have already started such lawsuits as a way to promote corporate responsibility. And the idea is, in fact, not so much to punish as to prevent damage before it is perpetrated. Late Polly Higgins' quest for making ecocide also a peace crime rather pointed to a more general framework of ecological justice.
A framework in which the planet is not seen as only an asset to be exploited for its resources, and in which blatant inequality is not furthered by the attack on the ecosystems that sustain and keep local communities safe.

Beyond the coming legal battles, therefore, it is perhaps more important to consider how an accepted idea of ecocide will promote new historical readings and prompts us to change our practices. Once the idea of ecocide reaches the heart of public discourse, it will surely bear retroactive effects in our appreciation of many everyday practices that followed from the Industrial Revolution. And this is one of the reasons why the notion of ecocide is a rhetorical game-changer within the rising field of climate justice.

As the notion of climate justice gives a new emphasis and direction to long lasting environmental struggles, so ecocide acquires new darker shades. This is no longer about the damage to a local ecosystem, the destruction of a specific biome, or the polluting of the resources in a given habitat. This is actually a global attack on interconnected planetary systems, with widespread, durable consequences — which first and foremost undermine those who are already the most vulnerable, be it the more impoverished human communities or species threatened by extinction. Climate justice brings ecocide to the fore at a much more impressive scale. And as climate justice broadens its reach, so ecocide will become more of a lens through which we look at practices that directly or indirectly prolong the impacts of climate change.

35. See See Robert Holtom, "Shell Case Echoes Call to Eradicate Ecocide," 22 Aug 2011. United Nations University. Accessed 17 May 2020, https://ourworld.unu.edu/en/shell-case-echoes-call-to-eradicate-ecocide

36. See Anja Gauger, Mai Pouye Rabatel-Fernel, Louise Kulbicki, Damien Short and Polly Higgins, The "Ecocide Project: Ecocide is the 5th missing Crime against Peace," Human Rights Consortium, University of London, 2012. Accessed 1 May 2020. https://ecocidelaw.com/wp-content/uploads/2019/06/Ecocide_research_report_19_July_13.pdf

In this context, it should not be a surprise — although it may still be — that climate justice is starting to make it into architecture and urban planning curricula. It certainly did at the Harvard Graduate School of Design, with the launch of a new course offer on the topic in the Fall semester of 2019. In the course's description, instructor Abby Spinak precisely highlights that recent discussions "increasingly make evident that climate change is much more than a technological problem of carbon mitigation." The climate crisis is also seen a symptom of structural social and economic challenges. Thus, those who want to explore the subject must "combine study of theories of justice, inequality, and structural violence with a deep dive into climate science, policy, and international diplomacy." And such an enquiry must then address diverse forms of activism, non-Western perspectives or the histories of geopolitical violence so as to disclose "what new kinds of practices, knowledges, and collaborations are necessary to build more just and responsible relationships between people and the nonhuman world, and with each other."

Following Spinak's thoughts in a related interview, such a take on climate justice ultimately stimulates designers and planners to demystify institutions and built environments as they exist today, providing "an opportunity to challenge epistemologies surrounding climate change." Perhaps only then can they bring "the social outcomes of ecological instability" to become a core motivation for their everyday practice. And in the scope of this book, this angle may prove more crucial.

If we are to clarify moral imperatives driving the evolution of architectural practices, their ability to integrate relevant social narratives is certainly key. Indeed, these conditions the capacity of architectural design to reach new stages in response to changing sensibilities. And sensibilities do change.

What was once proper in a context of progress may suddenly be reread as a unlawful contribute to a newly affirmed misdemeanour against society or a larger ecological system. Human slavery was once accepted as common, but it was redefined as utterly offensive under a new human rights perspective. Similarly, extractive practices and their beneficiaries will be deemed criminal under the light of a new environmental understanding. Similarly, architectural practices that were once accepted as inevitably driven by their structural context may too be revised under a new light of climate justice.

It is no wonder, then, that a film festival on architecture decided to choose "Guilty Architecture" as the theme for their edition of 2020. The 12th iteration of the Architecture Film Festival Rotterdam thus proposed their "spectacular selection of films and documentaries" to gravitate around how "devastating ecological consequences of global urbanization call on the moral understanding of developers and architects." This surely presupposed that there were sufficient projects going around to fuel an entire festival. Ultimately, the initiative served to leave participants and audiences with the suddenly actual, newly uncomfortable question of "how innocent is architecture."[37]

But also in the academic-related field, precursors of this new critical inquiry started to investigate the potential connections of the design disciplines to issues of climate justice. The research action of Forensic Architecture in this realm is exemplary. While most of their work has been dedicated to develop architectural tools that analyse and produce evidence of war crimes, they too realised that ecocide now deserved the application of their forensic tools. So, in between the thirteen categories of investigations that Forensic Architecture set for its guidelines, one is that of environmental violence. This is defined as part of a military strategy, or, otherwise, as the way in which any association or group can perform a destruction or reconfiguration of the built and natural environment and, thus, affect the "political subjectivity of native people and give rise to populations conducive to state or colonial control."[38]

In their early project on Guatemala's Ixil Triangle colonial massacre, Forensic Architecture deployed many of the digital tools that they would regularly expand from an architectural technical repertoire onto new forensic purposes. Data mining, geolocation,

37. See Architecture Film Festival Rotterdam, "Guilty Architecture," 6 March 2020. affr.com, Accessed 18 May 2020. https://www.affr.nl/en/festival-2020/theme-affr-2020-guilty-architecture/

38. See Elise Hunchuck, "Conditions of Life: Forensic Architecture in the Age of Climate Breakdown," in FlashArt #326, June-August 2019. Accessed 18 May 2020, https://flash---art.com/article/conditions-of-life-forensic-architecture-in-the-age-of-climate-breakdown/#_edn3

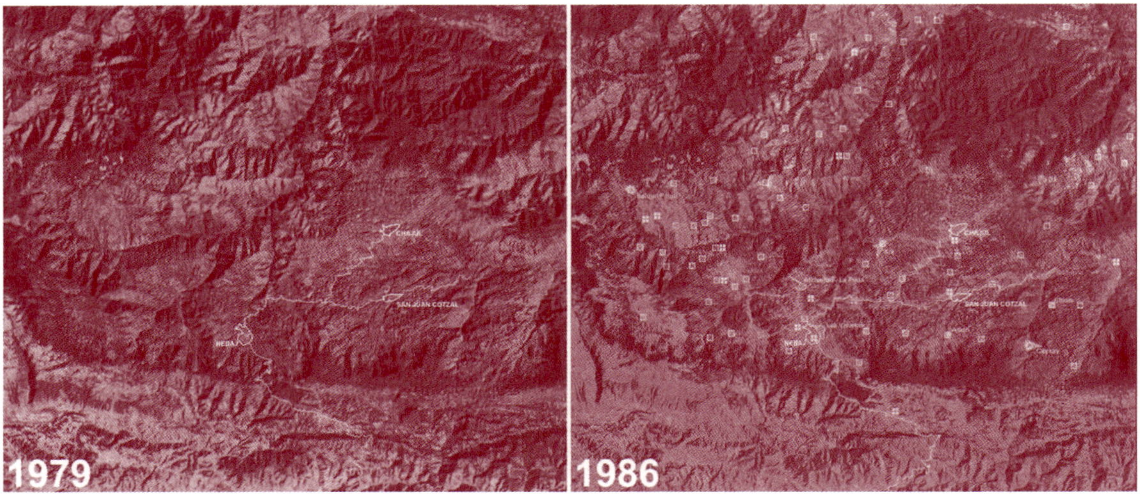

NDVI – 1979 to 1986 from Genocide in the Ixil Triangle, 2014. A Normalized Difference Vegetation Index (NDVI) analysis of the environmental transformation of the Ixil region between 1979 and 1986 shows substantial changes to the natural and urban environments in the region. © Forensic Architecture, 2021

pattern analysis, photogrammetry or 3D modelling were combined with other sophisticated engineering tools to analyse the local built environment. Through the medium of architecture, the agency accrued evidence of environmental aggressions to integrate a larger genocide trial against the state of Guatemala.

In several other projects, Forensic Architecture conducted investigations that ranged from forest fire ecocide in Indonesia[39] to herbicidal warfare in Palestine,[40] always providing architectural contributes to legal actions started by local communities, international NGOs and other activist organisations. Theirs is a radical take on how the very notion of architectural practice can undergo a metamorphosis as it reacts to the pressing summons of climate justice.

These are examples that reveal a shifting sense within the architecture community. Climate justice reveals itself as a magnifying device through which architectural responsibility can be reexamined. And yet, such forays at environmental fair play, and the transformations they may originate in the roles of architecture and urban planning, are still residual within the wider field of practice.

Before we dare to expect that the architectural profession at large changes its moral imperatives, we may have to go back to a broader historical reflection. We may have to dig deeper into the idea that Modern architecture was an inadvertent accomplice in

39. See Forensic Architecture, "Ecocide in Indonesia", 31 March 2017. forensic-architecture.org. Accessed 18 May 2020, https://forensic-architecture.org/investigation/ecocide-in-indonesia

40. See Forensic Architecture, "Herbicide Warfare in Gaza", 19 July 2019. forensic-architecture.org. Accessed 18 May 2020, https://forensic-architecture.org/investigation/herbicidal-warfare-in-gaza

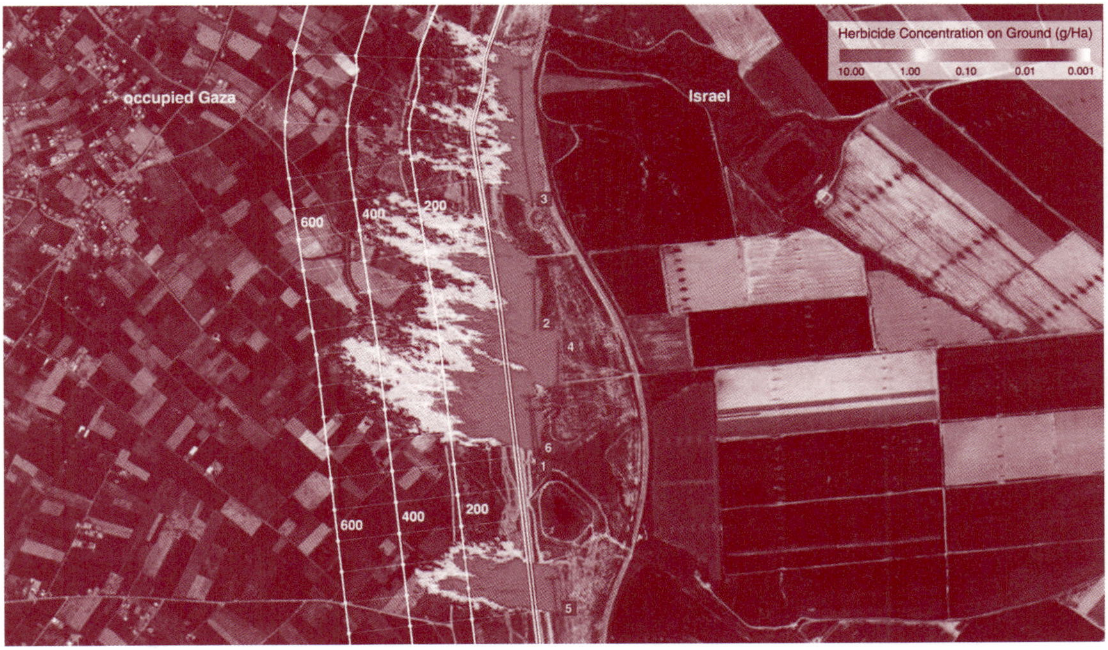

The results of Forensic Architecture's analysis show the distribution of concentration of herbicide as it travels westward into Gaza. (Forensic Architecture and Dr Salvador Navarro-Martinez) © Forensic Architecture, 2019

what, in the near future, will be considered as a continued practice of ecocide by industrialised human societies.

The last years of the 2010s were marked by an acceleration of the climate emergency awareness. This led to a visible increase of different forms of ecological protest and climate action. By 2019, for example, several cities in the US had launched their own lawsuits against oil companies. They were not entirely successful, but they certainly raised the question of how we are to address the most obvious culprits of a planetary environment gone rogue.

They also opened the way for attention to fall on other sensitive polluters. How long before the construction industry is also persecuted for its contribution for ecocide? Riding the predominant wave of politically correct historical revisionism, however, you could take this argument even further down the road. Architecture itself can be targeted. And not only for its current actions.

With their glorification of concrete, steel and glass, Modernist architects could be said to provide an enthusiastic boost for the embrace of ruinous environmental practices by the construction industry. With its rejection of adaptive vernacular practices, the International Style could be said to be a fellow conspirator in the progression of globalisation to abandon any concern for the balance of both local and planetary ecosystems. It may be argued that there were no evil intentions, or that the environmental consciousness that we today enjoy was absent. It may be held that architects were only celebrating the advancement of architecture through the available gifts of industrial progress. And yet, the ideological impact of that celebration must now be brought to question.

Who, if not architects, have been historical accomplices in producing an idealised desire for some of the favoured weapons of ecocide, such as the production of concrete or steel? The same historical accounts that built the praise of concrete, steel or glass as the ideal carriers of early modernist bravura, may now be reversed in the harshest of condemnations for a lack of ecological vision from the pioneers of 20th century architecture.

Starting a century ago, the massive adoption of reinforced concrete was a defining component of architectural modernity. As such, it provides a most poignant case of how the adoption of a material that is today seen as a tool of ecocide was elevated and reinforced by architectural discourse.

As overwhelming news pieces go, these days concrete is surely making the headlines. As a thorough essay in the British newspaper The Guardian pointed out in 2019, if concrete was a country, its carbon dioxide emissions would be "surpassed only by China and the US."[41] This is when alarms start to ring. When doomsday statistics tell us that every two years more concrete is produced than the 8bn tonnes of plastics made over six decades, our heads too start to ring. Despite its importance to global wellbeing, the material itself is suddenly seen as part of the weaponry through which the built environment has taken over nature.

Concrete's responsibility for 4 to 8% of the world's greenhouse gases, puts it right after coal, oil and gas in the most wanted list of ecocide felons. Which brings us to that "most

41. See Jonathan Watts, "Concrete, The Most Destructive Material on Earth," 25 February 2019. The Guardian, Accessed 9 Mar 2020, https://www.theguardian.com/cities/2019/feb/25/concrete-the-most-destructive-material-on-earth

severe, but least understood impact of concrete" mentioned in the Guardian text. That is that it "destroys natural infrastructure, without replacing the ecological functions that humanity depends on for fertilisation, pollination, flood control, oxygen production and water purification." Nonetheless, much before concrete was recently discovered by journalists as "the most destructive material on Earth," unknowingly architects and architectural commentators were the first and foremost to fetishize the material as a source of cultural rebirth.

As cultural advancement goes, indeed concrete became the most consumed substance across the globe after water. And it is true that this happened due to economic and developmental forces that go wildly beyond the influence of architects. But architecture's long-time promotion of concrete as a signifier of what it was to be modern certainly helped. From the early experiments of French architect Auguste Perret to Le Corbusier's genotypical Dom-ino structure — which historian William Curtis once described as an "image of origins, out of which a symbolic architecture might be developed"[42] — a consistent elevation of concrete by the most prestigious designers along the 20th century undoubtedly gave concrete legitimacy and status.

Concrete's solidity, malleability and ease of use were profusely illustrated in wide-ranging architectural experiments, making it appealing to a growing audience of builders, politicians, industrial leaders and, ultimately, clients and consumers. As Adrian Forty puts it in his book *Concrete, A Material History,* architects were central to the interpretation of concrete as a medium of culture. They helped shape the essentially modern assertion that "like the internal combustion engine, antibiotics, genetically modified crops and digital technology, concrete realized the prospect of transforming nature, and of transforming ourselves and our relationships with each other."[43]

Unlike previous historiographies of concrete as a leading protagonist of modern architecture, Forty's has the advantage of being already tinted by an ambiguity derived from an evolving

42. See William Curtis, in Modern Architecture Since 1900, 3rd Edition, London: Phaidon, 1996, p.85.

43. See Adrian Forty, Concrete and Culture: A Material History. London: Reaktion Books, 2012, ISBN 9781861898975

environmental perception. Writing in the early 2010's, he is still able to evoke how Walter Gropius's pictures of concrete silos and factories "fuelled the imaginations of the avant-garde." Like others before him, he can still remind us of how the drawings of Futurist Antonio Sant'Elia "encouraged the association of concrete and modernity" and reinforced the assumption that "concrete was a 'revolutionary' material." But he is also cautious in the way he recognises that concrete's expanded footprint in our society is a strong reason for the substance to be vilified — and, with growing proof, increasingly blamed for the erasure of nature. As if presenting a case in a slightly different light, he rather underlines contradictions. Aware of a changing narrative about the unsustainable and delinquent impact of certain building practices, he keeps a distance from the aura which architects have attributed to concrete along the past one hundred years.

This is a subtle shift in the classical historical appreciation of architects' bromance with concrete. It barely hints at the radical revisions that we may expect in a context of rising ecological emergency. The aggrandisement of concrete may not reach the courts of international justice, but its perpetrators will certainly be seen in a dimmer light. And this should remind us of the grim responsibilities that, inadvertently or not, architects have incurred — and how these could be an object of reflection and change. That architects may have been accomplices in ecocide should alarm us, and compel us to radically alter behaviours and objects of desire.

The same logic applies to other cherished fetishes of modernity and progress, which helped build the identity and culture of architecture along the 20th century. In the profession's grand narratives, concrete was Europe's or Japan's material of choice for both a new architecture, as well as the need to rebuild cheaply after World War II. In the United States, however, steel was the stuff that fed really dreams of upward growth. Soon, that idea too may rather sound like a nightmare. The production of steel, after all, comes close to concrete in the carbon dioxide villains chart. And glass, that symbol of modernity's conquering of transparency, also bears its opaque spots.

Antonio Sant'Elia, Italian, Part of the series La Città Nuova, 1914

As the "carbon issue" of Architect magazine made clear to their readers in the first days of 2020, steel production comes very close to concrete in its impressive environmental footprint.[44] Globally, this sector accounts to more than 10% of carbon dioxide emission. As it is the case for concrete, steel manufacturing is energy intensive, and a process that is difficult to decarbonise. Even if steel-making furnaces run on electricity, it is still to be discovered how renewable energies could provide their hungry needs. In the building sector, steel does enjoy the advantage that, unlike concrete, is highly recyclable, and about 98% of it gets re-used after demolition takes place. But until that happens, inclusively inside reinforced concrete, steel just adds to the embodied carbon that goes into the building's construction.

44. See Gideon Fink Shapiro, "Concrete, Steel, or Wood: Searching for Zero-Net-Carbon Structural Materials," 15 Jan 2020. architectmagazine.com, Accessed 18 May 2020, https://www.architectmagazine.com/technology/concrete-steel-or-wood-searching-for-zero-net-carbon-structural-materials_o

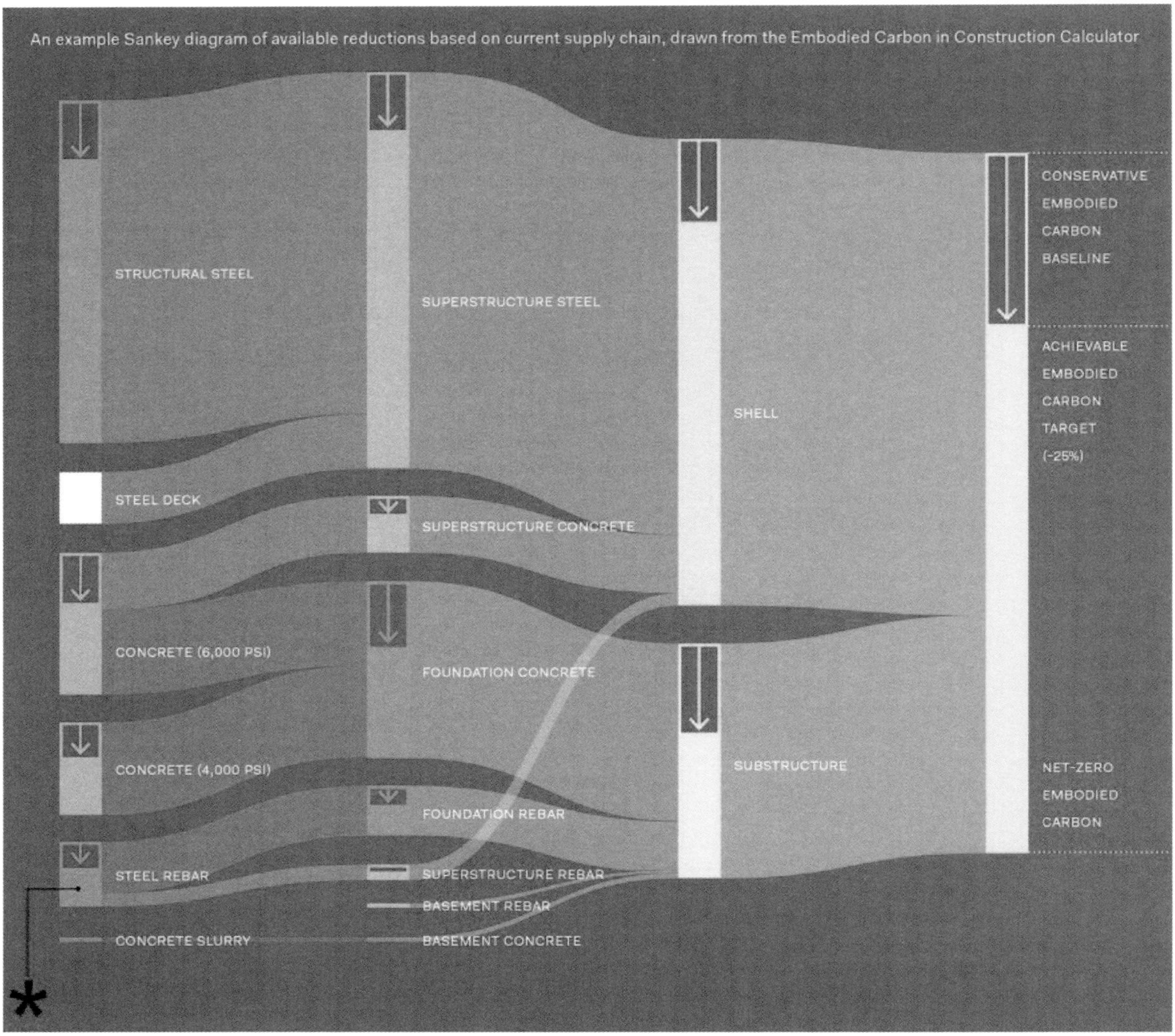

Sankey diagram of available reductions

Again, both these materials were central to grand narratives of modernist architecture, especially in its Miesian vein. Starting with London's Crystal Palace, structures in steel and glass became the most notable bequest of the Industrial Revolution towards modern architecture's need for a new visual language. Yet, the aestheticisation of steel, as well as the celebration of glass's transparency, were a justification for buildings which, besides their intrinsic, unacknowledged carbon footprint, also generally carried a reduction of energy efficiency.

The language of glass and steel was promoted as a stylistic must-have to be deployed in any climatic conditions across international borders. With this glorification came the pressure for architects to abandon more sensitive, context-wise responses to local climate conditions. And with the enthusiastic embrace of this progressive innovation along the 20th century, came the escalation in energy consumption, a multifold increase in air conditioning and, of course, greater unintended carbon dioxide emissions across the board.

The multiplying environmental impacts of steel and glass architecture were obviously out of sight for many decades. And it was only very recently that, in fact, they were targeted in unexpected policy stunts — as when New York City's Mayor Bill de Blasio loudly announced a local ban on steel and glass high-rises in mid-2019. As the New York Times quickly noted, it turned out that the measure intended to proscribe those skyscrapers that "have contributed so much to global warming," was not exactly a ban after all. Troubled "real-estate developers and architects" were soon relieved to know that this was, in fact, only a call for structures that would be "more energy-efficient." While de Blasio seemed to have spoken too soon on the sacred materials, it was true that the City Council had "passed legislation that would force buildings to reduce their greenhouse gas emissions by 40 percent by 2030."[45]

De Blasio's off-centre announcement did ring a different bell. Once climate populism takes over the susceptible minds of politicians, things may quickly change. And, under so-called climate dictatorship, what was taken for granted for many

45. See Jeffery C. Mays, De Blasio's 'Ban' on Glass and Steel Skyscrapers Isn't a Ban at All," 25 April 2019, New York Times. Accessed 18 May 2020, https://www.nytimes.com/2019/04/25/nyregion/glass-skyscraper-ban-nyc.html

decades, may suddenly be the object of a harsher scrutiny. So, with climate justice on the rise, and the awareness of the environmental emergency shifting to red in the radars of policy-makers, the moment may come in which the scope of ecocide is indeed extended to considerations of the built environment. This may not take the legal form of a new liability nightmare — for which, in any case, architects are already apparently having to prepare for.[46] But, in a more bewildering scenario, as the recent political response to the corona virus outbreak now oblige us to consider, this may well turn into an ecological witch hunt with unexpected effects.

46. See Larry Dany and Nick Boyd, "Are design professionals liable for failing to anticipate the effects of climate change?" 9 May 2020, archpaper.com, Accessed 18 May 2020, https://archpaper.com/2019/05/architect-liability-climate-change/

When one considers the retroactive targetings of ecocide, one should not be expecting that the early adopters of reinforced concrete, steel or glass are now vilified out of a sanctimonious history of architecture. We can rather try and understand from the past how the embrace of a new material or technology has the force to reshape the cultural identity of architecture, and subsequently of the built environment it helps produce. It is the excitement of welcoming new symbols and new motivations that has to be weaponised for alternative purposes.

Rather than welcoming the irruption of another formal autocracy in the realm of a more ecologically driven architecture practice, we can then just use greener materials and innovative carbon sequestering technologies to expand the possibilities of a new architectural language. And, importantly, this would not just mean other stylistic options. As Adrian Lahoud noted about the *Climate Crimes* installation with which we started this chapter, this could rather contribute to practices that reveal a new "empathy for the victims of an unthinking climate" — i.e. a new sense of responsibility towards a crisis in which we are all entangled, but from which some will suffer more than others.

As much as architecture culture is increasingly conscious of issues of inequality, gender, or racial discrimination, its discourse must also be progressively made more aware of and responsive to the complex implications of climate justice.

Reinforced concrete, steel and glass were once embraced as symbols of modernity and progress. Now, they may be seen as the aggressive ammunition with which the built environment continues to threaten climate justice and the planet's ecological system. And in that context, even if because of the risk of otherwise losing cultural significance, architects have to revise the palette of both their imperatives and their material choices to signify the willingness to enter what many are calling *The Great Transition*.

4.
The Great Transition

"Timber is the new concrete," the catchy mantra went.

An architect actually coined that slogan. In the same breadth, he justified that climate change awareness had entered "the domain of global emergency,"[47] and as such, timber was now a more relevant construction material than ever. Timber was to be the new object of architectural desire. With much fanfare, concrete and steel had taken over the heart of architecture and the construction industry along the 20th century. Now, it was the time for timber.

47. See Jason Sayer, "A new RIBA show asks if timber is the new concrete". 22 January 22, 2020. The Architect's Newspaper, Accessed 5 May 2020. https://archpaper.com/2020/01/a-new-riba-show-asks-if-timber-is-the-new-concrete/#gallery-0-slide-0

When the cofounder and partner of London-based dRMM architecture practice uttered the phrase in early 2020, he demonstrated a good sense of what was needed to start the transition of architecture culture to a new stage of environmentally friendly practice. Besides good ideas, that transition required two essential, often overlooked traits: media savviness and, on the other hand, a well-oiled promotion machine made of books, manifestoes, exhibitions, conferences, social media and whatever mass communication involves today.

As Le Corbusier had taught his peers one hundred years earlier, one needs those tools to turn new architectural concepts into the stuff of a wide-ranging cultural transformation. Faced with the sudden need to alter the carbon footprint of architectural practice, architects will require more than just good intentions. Being asked to be part of an urgent energetic transition towards a cleaner environment, they will need more than a style redo or another material palette. As brilliant as some unique ideas might be, they will have to be adopted *en masse*. Changing practices will need to enter a mainstream sensitivity. New positions will have to inspire some sort of herd behaviour.

Forest of Fabrication, the exhibition in the scope of which architect Alex de Rijke came up with that catchphrase, was a good example of initiatives that should be repeated *ad nauseam* — if, indeed, architects do want to be a significant part of the so-called *Great Transition*. The profession does need to be bombarded with viable alternatives to its current status-quo. It definitely needs stimuli to shift its acquired tastes.

From its title to its media reverberation, the otherwise modest monograph exhibition of 24 timber buildings managed to stir some conversation. It was also well-timed, as it opened around the period in which emergency alarms were ringing worldwide, and, simultaneously, timber buildings were popping up in architectural media as glimpses of hope for a less ecocide-prone architecture practice.

Held at the Royal Institute of British Architects' headquarters in Liverpool, the show had also enough of an institutional setting to make it a suitable vehicle for the debate of some broader issues. It suggested that, even if arising from specific architecture proposals, a different conversation could take place. As de Rijke made clear, this was an opportunity to discuss how sustainable forestry management would, in the near future, "be a really important step for how the world reduces its carbon footprint." His words hinted at how architects needed to go beyond formal navel-gazing and "look at the biodiversity element, the cultural element, the issue of land use." The exhibition pointed, after all, to ways in which architects could join *The Great Transition* through everyday, reachable practices.

Such aspirations were also being made palpable elsewhere. As mentioned above, narratives of promising timber structures were making the architectural news across the board. Generate Studio's CLT apartments, in Boston, for example, were widely publicised as "using a cross-laminated timber 'kit of parts' (…) that allows for disassembly and reuse."[48] Timber was explicitly defended as "less energy-intensive than conventional materials like concrete and steel," allowing the building to aim at net-zero-carbon operation. The case revealed a drive to innovate, with the MIT-taught architects soon announcing that their project had secured seed funding from the National Science Foundation. The ambition was made clear that, with buildings accounting for 40% of global greenhouse gas emissions, this research-based design offered an opportunity to "democratize mass timber systems."[49]

48. See Rima Sabina Aouf, "Generate and Placetailor to build carbon-neutral CLT apartment block in Boston,"14 Feb 2020. Dezeen, Accessed 6 May 2020. https://www.dezeen.com/2020/02/14/model-c-carbon-neutral-clt-apartment-block-boston-generate/

49. See "Generate Wins the National Science Foundation America's Seed Fund!", 29 April 2020. Generate Technologies, Accessed 6 May 2020, https://generate-technologies.com/news

Carbon-neutral CLT apartment block by studio Generate and developer Placetailor

Another project for Boston kept timber to its facades, but aligned with the city's plan to achieve carbon neutrality by 2050. As pledged by Toronto-based KPMB architects, when its 19 stories science centre would be completed in 2022, the new building for the University of Boston would to be the largest carbon-neutral structure in the region. Tellingly, it was also claimed that the building would be fossil fuel-free, relying on geothermal wells for its heating and cooling. Which meant it could thus be "designed without gas connections."[50]

50. See Katherine Guimapang, "KPMB Architect's new building for Boston University reaches for carbon neutrality". 5 Mar 2020. Archinect, Accessed 6 May 2020. https://archinect.com/news/article/150187626/kpmb-architect-s…

In the early 2020s, news like these created a positive vibe for environmental change in architecture. Yet, it would have to be said that most of these pledges were still only aimed at the carbon neutrality of the building's operation. Even when advocating to store carbon in their timber structure, those commitments did not necessarily address the totality of embodied carbon deployed at the time of the building's assembly.

Thus, when international players such as Oslo and New-York based Snohetta vowed to design only 'carbon-negative' buildings,

Boston University, Center for Computing & Data Sciences by KPMB architects

we knew we were entering a new stage of the decarbonisation game. The notion would imply that, over their lifetimes, Snohetta's projects would generate more energy than they would consume. As Kjetil Thorsen described it, "we have the ambition of having projects on the table that will become CO2 negative in the cradle-to-cradle definition." That is to say that, in time, those structures would also offset the 85% of carbon emissions that are present in the materials and processes used at the moment of their construction. While such pledge would involve heavy investment and dedicated research, it started to reveal a clearer understanding of the challenges presented by the *Great Transition*.

Indeed, given its slow start, the *Great Transition* will not only require reducing the rates at which we emit carbon dioxide. It will actually demand that we also take existing carbon out of the atmosphere. This is the reason why discussions on geo-engineering have been heating up. As scary as spreading sulphur dioxide in the stratosphere may sound, those and similar options are the ones currently available for research. But the risks and costs of carbon capture and storage also open opportunities for other forms of innovation. With the building stock being responsible for so much embodied carbon, one would expect that architecture too should join the bandwagon of the kind of environmental innovation that is indispensable for a successful decarbonisation.

As Naomi Klein hints in her seminal book *This Changes Everything*, it is not per chance that our transition to a decarbonised society is deemed *great*. It will certainly require a large amalgamation of small, reachable gestures. But it will also require more drastic and ambitious pledges. If we were to curb emissions in the decade after she was writing, we would need "a massive mobilization larger than any in history."[51]

51. See Naomi Klein, *This Changes Everything*, 2014, Simon & Schuster, NewYork: 2014. p.5

Klein was writing in 2014, so the best part of that decade has passed without barely any improvements. If so, global carbon dioxide emissions have grown worst. If she then decried that we needed "a Marshall Plan for the Earth," that plan is now ever more critical. And, if already at that time such precursor of coming Green New Deals needed to "mobilize financing and technology transfer on scales never seen before," you can predict how much more urgent it has become meanwhile. If, as Klein pleaded, there were only ten years to "get technology onto the ground in every country to ensure we reduce emissions while raising people's quality of life," you can start sensing the choking feeling of how great this transition must be.

But there is other meaning to greatness implicit in decarbonisation and an energy transition at global scale. That is the sense of opportunity. With radical necessities, come per force new ways of doing. As Klein also stresses, this is "a vision of the future that goes beyond just surviving or enduring climate change." Beyond mitigation or adaptation, as technical discourses go, this could be the sight of collectively levering a crisis to generate new social and economic benefits. This could represent the chance to stimulate technological advancements, but also to clasp new, sounder everyday practices. In Klein's words, "the resources required to rapidly move away from fossil fuels and prepare for the coming heavy weather could pull huge swaths of humanity out of poverty providing services now sorely lacking, from clean water to electricity."

This is the scenario in which architect's abilities are able to join a greater collective effort: using innovation in materials to revolutionise form-making; locating energy efficiency and sufficiency at the core of their conceptual thinking; imagining buildings that help produce resources instead of depleting them; replacing modernism's guilty pleasures with a new passion for the joys of decarbonization.

With Klein and others, architects must start discovering "all kinds of ways that climate change could become a catalysing force or positive change." Because, in the end, as this activist journalist lists, the *Great Transition* will imply rebuilding local economies, creating jobs in new sectors, rewrite trade deals, remaking agricultural systems, preparing for climate migration, investing in public infrastructure, and building a whole new lot of environmentally friendly affordable housing. Most of these are areas in which architecture has a contribute to give, whether it is just following demand, or, more interestingly, providing the proactive imagination of much-needed alternative solutions.

As author Elizabeth Stamp broke it down to the wider audience of the Architectural Digest magazine, circa 2020 the "architecture industry" was actually seeing climate impact as "a monumental opportunity." Even so, she did kick off with the reminder that "when it comes to climate change, building construction is a major culprit."[52] Opening quotes by Harvard GSD's Jesse Keenan warned that the architecture field was falling behind, even if "it thought that it had gotten to this penultimate moment of environmental stewardship and sustainability." But Stamp reports on the success stories of institutional commitments, industry goals translated into codes, as well as innovative examples of leadership in practice — such as BIG's Copen Hill, the

52. See Elizabeth Stamp, "How the Architecture Industry Is Reacting to Climate Change," 2 March 2020. Architectural Digest, Accessed 8 May 2020. https://www.architecturaldigest.com/story/climate-change-design-architecture

Copenhagen waste management facility cum ski slope that made decarbonisation look like a fun endeavour.

Tellingly, Stamp's story finishes on the high note that, when it comes to the need to transform the construction industry, ingenuity is one of architecture's "crucial weapons." Meaningful progress, as she recognises, imply "big ideas and a rejection of the *status quo*." Yet, in order to check fears of high costs of environmentally friendlier designs, it also implies that architects assume a pedagogical role in explaining their clients the long-term benefits, both in ecological and economic terms. Long-lasting assumptions in the economics of the built environment as-we-know-it must surely be reversed.

One of the challenges that Klein identifies in the *Great Transition* is that of embodying ideas of "selective degrowth" in everyday practices, where previous modern paradigms were based on the mirage of perpetual GDP growth. This, ideally, would also entail how "reducing our use of material resources could be managed in ways that actually improve quality of life overall." As we will detail in coming chapters, this is also where design disciplines may contribute with innovation in recycling techniques, in adopting greener materials, or in ideating other ways of operating cities. Where architectural discourse was once key to reinforce visions of modern development, it can now find new greatness in casting light on the demands of the coming transition.

One example of how this could happen, even if still in a modest version, comes from LA-based visualisation studio Kilograph in partnership with non-profit advocates Climate Cents. Hearing the call to "strengthen their involvement in fighting the climate crisis," they decided to focus in changing people's perceptions on how clean energy projects and infrastructure transformations could provide for a beneficial, appealing urban landscape. Rather than narrowing down on dire problems, they sought to inspire and give visible shape to positive aspirations. The collaborative thus hoped that they would "get people to realize that we don't have to settle for the carbon-based economy as we know it."[53]

53. See Katherine Guimapang, "Kilograph and Climate Cents Team Up To Rewrite L.A.'s Climate Action Narrative." Archinect, Accessed 6 May 2020. https://archinect.com/features/article/150176134/kilograph-and-climate-cents-team-up-to-rewrite-l-a-s-climate-action-narrative?

CopenHill, Copenhagen
2019 by BIG.
Photo by Rasmus Hjortshoj

The sexy, fictional visualizations of potential green futures made by Kilograph and Climate Cents are significant. They are an important part of the myriad of actions that we need to see emerging in the architecture field in response to the *Great Transition*. But they also raise some concerns, which must be addressed right now. We must ponder that, while these presentations stir the debate and provide a temporary good feel, they may also lead us to avoid tougher decisions on how to effectively change architectural practice. As important as they are, while we focus on communication aspects, we risk postponing a deeper reflection on how architecture itself must transform and adapt.

By focusing on infrastructure — with vast amounts of territorial displays of solar panels — these alluring images could subtly drive us away from the idea that architectural form itself can contribute towards climate action. Such illustrations could underline a message that the architectural profession has been unconsciously amassing over the last few years. They could imply that climate action was essentially a technical, infrastructural, at most urban planning trope. And that, by welcoming such external inputs, we could continue to excuse buildings that are still engaged with massive, ecocidal carbon footprints.

As that collaboration between visualisation studio and climate action advocates, the architecture field will need many other alliances if it wants to envision its own transition beyond a now narrow disciplinary heritage. It will need to partner again with technological innovators, but so as to let technical ideas effectively propel its spatial conceptions. It will need to confer with ecologists and environmental engineers, but so as imagine how to turn buildings into living, productive entities. And it will have to learn from their close colleagues, landscape architects, not to add on some last-minute green features, but to devise how natural carbon drawing strategies can become a permanent feature of their architectural proposals.

To better understand the scope and variety of those strategies architectural, practices may always turn to today's decarbonisation bible: Paul Hawkens' *Drawdown*, a.k.a. "The Most Comprehensive Plan Ever Proposed to Reverse Global Warming." For this project, which also includes a complete website where the scientific research of many voluntary contributors continues to add up, Hawken established an ambitious goal regarding the drawdown of greenhouse gases from the planet's atmosphere. Out of curiosity, rather than anxiety, he proposed "to identify, measure, and model one hundred substantive solutions to determine how much we could accomplish within three decades towards that end."[54]

54. See Paul Hawken, *Drawdown*, Penguin Books, New York: 2017.

Solutions analysed in *Drawdown*, including how many gigatons of carbon dioxide they are expected to draw from the atmosphere until 2050, range from renewable energies to food systems, from transport to land use, and from materials to correcting gender inequalities. When one sees such an extensive roadmap for decarbonization, one soon realises that many of the pathways identified in *Drawdown* surpass the disciplinary competences of architecture. Initially, it may be hard to grasp how composting or producing biochar relate to architecture. But once one puts her or his head to it, it might be surprising how connections start to flourish. To have large scale impact, many of these techniques are industrialised. But, in several cases, they can also be distributed through individual or community nodes — in ways that end up relating to urban planning or architectural conception. The obvious examples here are solar or wind energy: they may come in extensive farms, but in their more domestic versions they can also integrate rooftops or even your tall building design.

Apart from entries on alternative cement, industrial recycling, refrigeration or air conditioning, a whole chapter of Hawken's research is, naturally, dedicated to buildings in cities. Anticipating that cities may come full circle — "from blaming them for environmental destruction to considering that urban environments, properly designed and managed, can be a kind of biological as well as a cultural ark"— *Drawdown* lists an array of solutions that *should* deeply affect architecture creation. These include the usual suspects, as green rooftops, net zero buildings and retrofits with better insulation or energy-reducing technologies such as LED lighting, heat pumps, smart glass, and automation in general. But they also reach to the transformations implied by bike infrastructures, autonomous vehicles, walkable cities, or greener waste and water management.

As for the timber construction with which we started this chapter, it wins a special place in *Drawdown*'s section dedicated to *Coming Attractions*. These are the emerging technologies that hold more promise to help reduce carbon footprints in the near future. There, it joins so-called "living buildings," which might still come across as science fiction. In fact, however, the expression designates a new certification system that, rather than sustainability of construction processes, values the regenerative capacity with which architectural structures can "reanimate and renew the environment, for both the natural world and human communities."

As Hawken's team informs us in addition to this superb panoply of urban-related drawdown techniques, it is no wonder that there are already biologists who are "mapping how cities could be more productive in terms of air, water, flora, fauna, pollinators, and carbon sequestration than the original lands they were built upon." From truly vertical farms to the layering of drawdown techniques that is possible in a single building, urban densification can be beneficial in more ways than usually expected. From regenerative agriculture, we can next move forward to regenerative architecture.

Many of the decarbonisation pathways quantified in *Drawdown* do involve architecture, buildings and urban infrastructure. But, as indicated above, there is also no reason for architects not to find unexpected inspiration even in techniques and concepts that are further away from their field. Even the less likely avenues can open up new chances for innovation and design ingenuity.

Over the last decades, even if often superficially, architecture culture has been able to take inspiration from the literary field, from critical theory, from biomimicry, from weather patterns and algorithms, from feminism and post colonialism. The list is apparently infinite. So, why would we not be eager to embrace the teachings of climate change? Why wouldn't we now be learning from the environmental crisis?

As I've hinted at earlier, architecture's *Great Transition* over the coming years also consists in shifting where its kicks derive from. We architects need to change the locus of our disciplinary climaxes. We need to reconsider our objects of desire.

As such, in the face of the radical change in social, economic and political behaviours required by the *Great Transition*, some inexorable questions again come to mind. How much are architects ready to embrace new modes of thinking about their built structures as part of a larger ecosystem? How will architecture practice effectively move beyond the murderous modernist paradigms described in the previous chapter? How much is architecture culture willing to have a second stab at being truly postmodern?

Because, let us face it: in ecological terms, postmodern architecture has never really happened.

From an environmental point of view, architects have never moved beyond modernising mantras. Even if we questioned modernism's formal rules, and briefly replaced their shapes and vocabulary, we have never moved beyond the superstructural aspects of the International Style. As hardly as Marxist theorist Frederic Jameson may have argued for it, architects have never addressed the intrinsic economic web of dependencies of architecture on a mercenary, global market. They have never dug deep enough into notions of growth and development underlying high modernism.

As much as, at some time, we thought we were learning something anew from popular culture, we were only further imbedding architecture in an unsustainable logic of predatory consumption. The famous adage introduced in 1972 by Robert Venturi, Denise Scott-Brown and Steven Izenour stated we should "learn from Las Vegas." But, this was actually only meant to make highbrow architecture more appealing to the voracious appetite of popular consumers. As poisonous concrete moved from a structural upholder of modernization to added painted decoration, postmodernism proved to be only a cheaper, more disposable form of the same conspicuous modern productive apparatus.

Postmodernist forms of architecture were still entangled, if not further enmeshed, in capitalist modes. If there had been experiments or attempts to leave a modernist paradigm behind, as we have analysed in the case of the same period's environmental avant-gardes, those remained lateral and unfulfilled. When it came to overcome modernism, the catchier post-modernist movement was undeserving of any revolutionary labels, indeed proving to be basically reactionary.

The *Great Transition* may be the one wider societal movement that, actually, provides the right motivation for architecture culture to leave behind the most criminal aspects of its modernist legacy. It may be what it takes to force architects to acknowledge that it takes more than cosmetic actions to surpass a fossil fuel driven modernity. And notably, younger generations in the field of architectural academic discourse are reaching, and arguing, for this consciousness. They, too, are discovering that if architects are to be a significant part of the *Great Transition*, they must first deal with a rather toxic past.

In the opening statements of Log magazine's special edition on "the death of carbon modernity," co-editor Elisa Iturbe puts the finger on the wound. "Given architecture's current entanglement with carbon capitalism," she states, the discipline "is clearly in need of a new ambition, creating an unlikely parallel to the Modern Movement."[55] The difference, she claims, is that when Modernists arose, they surfed an energy transition that had already happened. They helped construct the benefits of a new industrial boom fuelled by carbon. Now, however, a new massive energy transition is still ahead of us. And this means architects can help thinking through its first steps, and also actively contribute to shape its spatial configurations.

55. See Elisa Iturbe, "Architecture and the Death of Carbon Modernity", pp.11-22, Log #47, Anyone Corporation, New York, 2020.

The *Great Transition* is all about moving towards the clean forms of energy that are needed for decarbonised, but still economically functioning society. Iturbe rightly notes that,

Log 47: Overcoming Carbon Form (Fall 2019), a thematic issue guest edited by Elisa Iturbe

beyond immediate technical considerations, such energy must be taken "as a political and cultural force with inevitable spatial repercussions." That implicates architecture in its demeanours to conceive the built environment looking forward. Where I've earlier pointed to complicity in ecocide, Iturbe adds that such recognition indeed implicates "architectural and urban form in the creation and unfolding of the anthropogenic climate crisis." And this happens in ways that go beyond the easier path of an "immediate quantification of a carbon footprint."

Having architecture join the *Great Transition* is not simply an emissions accounting task, like those implicit in LEED-type certifications. It means assuming a greater responsibility, which Iturbe correctly posits beyond the typical "increases in energy efficiency and reductions in the carbon emissions of individual buildings." As she reiterates the uncomfortable truth, it is indeed inevitable that, if dependent on a carbon economy, "the built environment as we know it will be fundamentally unable to supplant the current energy paradigm or to address the climate crisis."

Re-establishing the comparison with the early 20th century avant-gardes, the question remains if architects are able to find again the power to "give form to emergent social structures" — with particular focus on the requirements of the *Great Transition*. As Iturbe concludes, "architecture's ability to transform the city in our time will not result from the incorporation of technological advancements in building systems, fabrication, or design software." And as I would certainly concur, this depends "on how radically architects are willing to reject the carbon fuelled myth of unlimited growth and lay the ground work for the coming transitions in energy and social form."

As a society, we still interrogate our ability to adopt a sweeping energy transition without fundamentally disrupting the extractive nature of neoliberal economies. As they say, it seems easier to destroy the planet than to replace capitalism. It is no wonder that the subtitle of Naomi Klein's book is "Capitalism vs. The Climate." In that battle, she notes, "we are failing to rise to the climate moment." Reasons for defeat are that the actions required by a *Great Transition* challenge many of the grand narratives "on which Western cultures are founded." These include the myth of our superiority to Nature or the notion of consumption as an integral part of our identities. But, above all, the great trial is to defy "our reigning economic paradigm"— the current normative that the author spells out as a mix of deregulated capitalism, public austerity and powerful private interests from a small elite.

These are the grounds on which, despite a huge "risk of destroying lives," climate change has really never "received the crisis treatment from our leaders." Rapid measures were taken in the face of the 2008 financial meltdown or, more recently, the expansion of the corona virus pandemic. However, policy-makers have been slow to assume the urgency of the *Great Transition* as an essential mechanism to slow down our current rush towards the so-called 6th Mass Extinction.[56]

56. See Elizabeth Kolbert, *The Sixth Extinction: An Unnatural History*, Henry Holt and Co. New York: 2014.

57. See Editorial Board, "Architecture, Design and Behavioural Science Need to Get Talking about Sustainability." 10 December 2019. Nature Magazine, Issue 576, 2019. Assessed 6 May 2020. https://www.nature.com/articles/

For a shift in this suicidal course, Klein suggests, it will take many of us to "stop looking away and decide that climate change is a crisis worthy of Marshall Plan levels of response." As it happened with the impact of civic movements in historical rejections of apartheid or sexual discrimination, politicians will then take note and act on it. They will make the necessary resources available for the *Great Transition*. And as the recent Covid 19 outbreak has shown, they will even prove Klein right in the prediction that they will also bend "the free market rules that have proven so pliable when elite interests are in peril."

Moving beyond the prevailing economic system of capitalism may be an existential issue that architects are incapable of addressing, much less through form-making. It is an issue that previously requires a wider societal debate on how to change business-as-usual. But it is within that debate that one can also ask how to disentangle architecture practice from its deep ties to the still prevailing logic of imperative economic growth. Then, as some specialists have already proposed in the context of the climate crisis,[57] the discussion may be brought to question the behavioural *status quo* of the architecture world. And seemingly, behaviours can indeed change overnight.

Bracing against the recent corona virus pandemic, we have seen plenty of surprising behavioural gear-changing from governments and individuals. As never before, we have briefly suspended capitalist norms of production and consumption for the sake of saving lives — with tremendous economic consequences. This has shown that, when required, it is possible to start changes that were deemed improbable, if not impossible, only a few weeks before.

Many have also posited the possibility that the ensuing economic recovery could be designed according to new rules. Despite an apparent desperation to go back to normal, people have realised that this might prove to be unlikely. The pandemic crisis has underlined the inequalities and uncertainties of the previous status-quo. Bailing out polluting companies, such as airlines or fossil fuel industries, could come with strings attached. Having had a glimpse of immobilised, but less polluted cities across the planet, the "new normal" could entail aspirations for a more balanced environment.

For optimists, the pandemic could actually turn to be one trigger for the *Great Transition*. It could provide the excuse to launch measures that were being hotly debated just before the pandemic hit. In spite of pressures to heat the economy, even by shockingly dropping environmental constraints, as China and the United States promptly did, the opportunity could arise for the political discussion of a Green New Deal to get a new impulse and traction.

As the environment editor from The Economist magazine has put it during the Covid 19 crisis, "the pandemic will end." It will leave memories of how, when summoned, countries and people did what it took to contain it. For sure, it will leave some hard lessons on how radical systemic changes look like if we suddenly need to reverse climate change. Ultimately, the pandemic has suggested, that, when it comes to the ecological crisis, and once that emergency too is well perceived, "the sooner societies begin to adapt and cut emissions, the less it will hurt."[58]

Changes in the way we conceive and tend to the built environment can certainly avoid a lot of pain.

58. See Catherine Brahic, The Climate Issue, 6 April 2020. The Economist, Accessed 6 May 2020. https://view.e.economist.

5.
Thinking Like a Building

In a somewhat surrealist dialogue, an oversized fly amanita mushroom tries to convince an old stone wall that it is time to let go. The time has arrived for the centuries old fortification to accept the final stages of its decay, to let itself crumble apart, and to enter the stage of a more complex web of interconnected organisms. Although it recognises it has already changed, and it is already partially made of other things, the obsolete rampart resists to lose its identity. The wall refuses to abandon its symbolic representation of human permanence. It refuses to let go of its Vitruvian *firmitas* so as to embrace a bigger-than-life fluidity.

The unlikely interspecies conversation is the tour de force of a 2018 film work by Danish art collective Superflex. In a mix of magic realism, fake 1970s documentary footage, and Monty Python comic overtones, the video lays out a compelling worldview of cyclic forces determining the course of Nature. As the authors describe it, *Western Rampart* challenges "our perception of borders and boundaries, whether natural or human built, through time."[59]

59. See Superflex, Western Rampart, 2018. Accessed 12 May 2020. https://superflex.net/tools/western_rampart

Western Rampart by SUPERFLEX

The film brings out two contrasting existential standpoints. In one, architecture stands for the desire for borders and immobility. In the other, the mushroom character "advocates for constant motion and circulation, as everything is meant to continuously transform to thrive." Even if by way of irony, this opposition invites us to question how human architectural endeavours are taken for granted as inorganic, anti-natural structures.

What if we could think of the built environment in a different way? What if we could reimagine architecture and other built infrastructures as part of that organic web that we call the natural environment?

Actually, this is the essential proposition of Steven Vogel's book *Thinking Like a Mall.* Within a broader discussion of environmental philosophy today, Vogel too points to a need to overcome obsolete distinctions between nature and man-made environments. "To be concerned with the protection of nature, under conditions of modern technological development," Vogel starts, "is inevitably to worry it might be too late, that nature might already have ended."[60]

We may have to overcome traditional assertions of "protecting nature," which forgo that we are now mostly enveloped by "the built environment of cities, or the technological infrastructure that modernity seems to require." With the overarching effects of climate change, it would be hard to find a single patch of the planet that has not been made artificial by the impacts of human technologies. As previously equated in the writings of environmentalist Bill McKibben[61] or philosopher Eric Katz,[62] we may conclude that nature is now man-made, and even operations of ecological restoration could now be labelled as just another human artefact.

Hence, in its "philosophy after the end of nature," as the book's subtitle goes, Vogel wonders if the time has not come for an environmentalism of the built environment. As this author warns, to persist in narrow notions such as those of wilderness protection might simply mean to "deemphasize, or even to ignore, the problem of determining a way for human beings to live in an environment in a sustainable and ecologically healthy

60. See Steven Vogel, *Thinking Like a Mall*, MIT Press, Cambridge, London: 2015

61. See Bill McKibben, *The End of Nature*, Random House, New York:1989

62. See Eric Katz, Nature as Subject, Rowman & Littlefield, Lanham:1997

manner." And for this holistic necessity to become a common sense, everyday endeavour, we must then enquire on buildings and urban features as integral to one environmental reality. We must understand what is it to be "thinking like a mall" or, for that sake, what is it to be "thinking as an old rampart."

In his quest for the environmental ethic of a "postnatural" world, Vogel insists in rejecting the distinction between nature and the human-made. This rejection could help push our current alienation from the environment to be substituted by new forms of responsibility in practices of the built environment. As Vogel takes from Karl Marx, "we are alienated from our environment when we fail to recognise it as the product of our own actions and thus fail to acknowledge our own responsibility for it."

When the life-cycle of buildings, or the consequences of urban sprawl due to market forces, are seen as naturalised, i.e. "as natural facts about which there is nothing we can do," the very same environmental alienation occurs. This can be counteracted first by acknowledging alienation itself. But secondly, it can be made transformative by addressing the products of human intention as an environment for which we care.

Vogel's idea of "thinking like a mall" is inspired by a passage from a 1949 book by American ecologist Aldo Leopold. In *A Sand County Almanac,* Leopold muses on a personal episode at a mountain range where he used to hunt. Years later, he gathers the unintended consequences of human prey upon the animals of that ecosystem, and concludes that humans must learn to "think as a mountain." Only then will they will start to grasp the delicate balance of ecosystems. Only then they will start "to see that there is more to the world than we understand, and to recognise the dark complexity and depth of the processes of nature that so exceed our limited ability to grasp and to control."

In this light, "thinking like a mall" means to accept that also objects and structures produced by humans have their own independent processes, which we not always anticipate and control. Evoking the story of a failed mall in his hometown, Vogel reminds us that, from the very moment of its construction, "the mall began slowly to crumble — as all buildings do, as all living things do." So, developing an empathy with its story might engage us with the similarities that the life of that mall bears with natural processes.

If we "think like a mall," the lesson goes, we can better understand the place of human artefacts within a system that is larger-than-human. By accepting the mall or the stone fortification as the "living buildings" mentioned in the previous chapter, we can start conceiving how they best fit and potentially enhance the overall environment.

We can eventually discover the life cycles of buildings as contributing to ecological processes, as much as we can understand them as providers of so-called ecosystem services. And as a climate emergency economy starts to account for the money value of those ecosystem services, we can investigate and invest in the right options for buildings and cities to become actual ecological tools. As mentioned earlier, buildings may help decarbonise, but they can also contribute in other, unexpected manners.

We can interrogate cities, factories, the obsolete mall or the obsolete fortress so as to understand how they fit into a wider, enmeshed ecosystem. As Vogel adds, we have a choice regarding what we call "nature," and we can address those same entities with the question we have once addressed at Nature: "Is this the sort of world we want to be environed by?" As that primal Nature was utterly transformed by human intention, so too the built environment can be managed to better respond to the needs of a more sustainable ecological system. That is to say, the built environment does not really have to be puppeteered by the invisible, nature-like hand of the market.

With "insights from fields such as city planning, urban aesthetics, architecture, and the psychology and sociology of place and space — but also," Vogel continues, "from ecology, biology, and restoration ecology," the built environment could become a more ambitious political project. Before it would be inadvertently returned to nature as a newly inhabited ruin — as author Alan

Island House in Laguna Grande. Andrés Jaque / Office for Political Innovation with Patrick Craine. Corpus Chirsti, Texas

63. See Alan Weisman, *The World Without Us*, St. Martin's Press, New York: 2007

Weisman has poetically depicted it in *The World Without Us* [63]—the environment we build could start being designed as part of a larger set of problems and balances. It could start responding to the needs of not only humans, but also of other species with which we share our surroundings.

Notably, the concern with interspecies relationships within a broader understanding of the human-managed environment is one that, under the influence of contemporary art practices, finally started to make its way into the architectural field. The notion is now the topic for still vaguely incipient academic theses, but also for budding experiments on what future architecture could look like. But, more importantly it is starting to inform environmentally-concerned practices into a new potential avant-garde momentum.

An interesting example of such direction was included in the international exhibition project *Eco-Visionaries*. In the words of the project's authors, the *Island House* project for Texas' Laguna Grande, in Corpus Christi, is conceived as "a device that empowers the environmental diversity" of its location. As Andrés Jaque and the Office for Political Innovation describe it, the dwelling acts as "an environmental mediator:" it collects and stores rainwater throughout the year so it may be used to automatically counter high toxicity and salinity levels produced by nearby polluting oil industries. In this, the architecture caters not only for humans, but also for the basic living needs of existing flora and fauna. With its conceptual strategy, the design addresses the fundamental issue of "how a small-scale project can be related to processes and challenges in the bigger context of climate change." Yet, more importantly, the concept also "provides evidence of the potential of a transition to new collective environmental and interspecies alliances."[64]

64. See Barry Samaha, "The Royal Academy of Arts Continues the Conversation on Environmental Protection," 22 August 2019. Surface Magazine, Accessed 25 Feb 2020. <https://www.surfacemag.com/articles/royal-academy-art-eco-visionaries/>

Indeed, following on Vogel's arguments, architects can stop considering buildings and urban structures as entities that are excluded and separated from a broader understanding of the environment. Architects and urbanists could find inspiration in the idea that, rather than continue to surgically separate architecture and urban structures from natural processes, we should eliminate such conceptual and physical boundaries and embrace a new hybridity. And as much as Vogel defends that we can embrace an environmentalism of the built environment, we too can start sketching a unified theory of practice where architecture is again seen as an organic part of a wider ecological system.

We still tend to think of architecture as one of those human artefacts that separate us from nature. From earlier notions of shelter, and the following decorated shed, we do insist in increasingly remit architecture to the realm of technologies that distance us from a pristine, untouched natural world. It will take yet another sort of conceptual climax change to make us move beyond that paradigm.

When it comes to the quickly expanding ecological damage engulfing us, such conceptual shift can be very useful in extracting architecture out of its inconvenient position as an accomplice of ecocide. It can help architecture culture move from the wrong side of the fence where it now sits.

This perceptual shift has, of course, happened in certain moments of history. From the Zen embrace of ephemerality in traditional Japanese architecture, to successive movements of organic architecture, architects have occasionally played with the idea of constructions that mimic, learn from, or think with Nature.

In the first instance, ancient Japanese dwellings were made in the acceptance that certain natural conditions were to remain unmastered by humans. In the face of recurrent weather disasters, structures were lightweight and ready to be quickly reassembled when blown away by heavy storms. This respect of processes of nature that exceed our control nevertheless originated a highly aestheticised and fulfilling experience of architecture. Traditional Japanese architecture is also beautiful, and remains influential to this day because of its intimate connection to Nature's cycles.

In the other case, continued Western ideas of organic architecture along the 19th and early 20th centuries suggested different pathways to establish formal and structural parallels to natural elements. From a decorative style inspired by natural forms, organic architecture evolved to effective attempts at the incorporation of Nature in architectural design, be it with Antoni Gaudi's embodiment of natural laws or Frank Lloyd Wright's blending of buildings into their surrounding context. By the end of the 20th century, biomimicry enhanced by digital capabilities brought this tendency to its maximum formal expressions. And yet, these experiments remain strangely disconnected from a truly ecological take on "living buildings."

Trends of "organic architecture", as well as more or less scientific takes on "bioarchitecture," are still around today as a continuous mishmash of sustainability, contextual care and do-good intentions. Yet, most of these generally remain frustrating in both their lacking design quality and, more alarmingly, their feeble responses to a wider environmental crisis.

For bolder principles on this front, one would probably have to again revisit those margins of practice that have been evoked here as the environmental avant-gardes of the 1970s. As one this period's protagonists told me, although marginal those early environmental experiments were already pursuing more radical features than those we see today under the banner

Earthsip Biotecture
by Mike Reynolds

of sustainability. As Mike Reynolds put it, "it was unheard of that architecture would address the six basic needs of humanity to survive, as we had identified in analogy to a spaceship or a vessel."

This was precisely what his "biotecture" was pursuing: to be a vessel that provided for "water, power, food production, on-site sewage treatment, recycling of your own waste and comfortable shelter that did not require fossil fuels." In the eye of these visionaries, it was already the case that in the future, more than a pretty box connected to the grid, "a building will be its own utilities."

As Reynolds added, "now that we know how to do it, there is nothing stopping us, there is no excuse for us not to go in this direction now." For this architect, after 50 years of pursuing technical solutions to his ecological quest, it was clear that we no longer have time "to play with rhetoric for a couple of decades, and then do something." It is pretty clear that whatever has to be done, "we have to do it tomorrow."

And indeed, lately, the pressing need to decarbonise has opened the way to go back to such perspectives. Together with emergent technologies, architecture can also accommodate those "biotecture" elements that will provide for some sort of tangible ecosystem service. The discussion on how architecture can collect rain water, produce renewable energy, or recycle waste products is back on the table, together with many new options made viable by a more sophisticated technology.

And, of course, these options also include the pamphletarian, if sometimes decorative, inclusion of generous amounts of 'green' in many recent architectural projects. However, these may also turn to be serious case studies on how to newly achieve a multi-layered, biologically productive city.

In one much publicised and awarded high-rise finished in 2014, Italian architect Stefano Boeri proposed his *Vertical Forest* as a "prototype building for a new format of architectural biodiversity."[65] Filled with 800 trees, 15,000 perennial plants and 5,000 shrubs, this residential tower in central Milan is said to provide in only 3,000 sqm of urban surface the equivalent vegetation of 30,000 sqm of woodland. And this produces particular ecological benefits: less warming from solar reflection; the generation of a microclimate that regulates humidity, produces oxygen and absorbs CO2 and pollution; and, last but not least, an inviting habitat for bird species. After a three-year long collaboration with botanists and ethologists, these characteristics were notably honed into "the architectural language and its expressive qualities" and into the building's primary conceptual strategy. With its balconies turned to protruding holding platforms, the dematerialised and iconic tower becomes, in the words of the architects, "a home for trees that also houses humans and birds."

65. See Stefano Boeri, "Vertical Forest." Stefano Boeri Architetti, Accessed 13 May 2020. https://www.stefanoboeriarchitetti.net/en/project/vertical-forest/

The *Vertical Forest* exemplifies how ecological thinking can successfully fuse with architecture's functional needs and aesthetic considerations. In a context of ecological pressure for decarbonization, it generates a meaningful paradigm prone to be emulated and reproduced in different contexts. It also illustrates how the

Milan Vertical Forest
by Boeri Studio,
Image by The Blink Fish

expanded notion of "thinking like a building" can help us surpass a limited, non-productive understanding of architectural artefacts.

Thus, the architects of the *Vertical Forest* present it not so much as "a simple architectural object," but rather as "set of processes — partly natural, partly man-managed — that accompany the life and growth of the inhabited organism over time." As it would be expected, attending to the needs of the plants, as well as to the "control of the anthropic-vegetal balance," become central everyday tasks in the building's maintenance. From the automated irrigation to greening operations, as seasons pass the architectural narrative continues beyond the completion and delivery of the built structure. In its daily functioning, the building provides for a gradual recolonisation of the city by natural organisms.

Building stocks will necessarily have to be replaced as cities, too, go over seasons. New needs are always on the horizon. Old buildings decay and give way to new constructions. In a context of climate emergency one would thus expect that, soon, it will be mandatory that any new buildings will offer a thorough palette of ecological services — as those that the *Vertical Forest* hints at. This would be a genuine regulatory revolution. As we have glimpsed in the previous chapter, its decarbonisation potential would have a vast environmental impact. And with ecological catastrophe knocking at our doors, it shows us the way to more responsible practices.

As Steven Vogel pointed out, thinking with and of buildings as part of an environmental totality is an important step in overcoming our alienation from broader ecological issues. Particularly for architects and urban planners, it will certainly help us recognise that "the environment comes to be what it is through our practices" and, hence, that we can also "engage in different practices."[66]

66. See Steven Vogel, idem, p.92

As the author of *Thinking Like a Mall* puts it, we can "genuinely acknowledge our active involvement in transforming the world only through action." For architectural practice, as for other activities that generate a carbon impact, that means self-consciousness about their own significance, i.e., a more acute awareness that "they do transform the environment." As Vogel suggests, we dwell in the assumption that we can alter the laws of our social and built environment — from its economics to its political norms — as little as we would be able to twist "the laws of gravity or of thermodynamics." Even if without a deliberate intention, that assumption allow us to collectively produce much ecological harm in what some call "a tragedy of the commons." The expanded environmentalism that this author calls for, ultimately arriving at a larger democratic participation, thus firstly involves practices "in which the actors acknowledge and take communal responsibility for their transformative effects on the world."

For these reasons, it is essential to consider a systemic approach to the environmental roles of the artefacts that arise from the field of architecture. And this approach should not only affect the conception of buildings, but also urgently extend to the idealisation of urban infrastructure and cities in constant need of retrofit.

As lately it has become manifest, the conventional grey infrastructure of cities — made out of concrete, asphalt and other toxic materials — also needs to be urgently substituted by its greener equivalents, in what is usually called ecological urban design.[67] Examples range from the green and facades mentioned earlier, to rain gardens, urban farms, linear parks and many others. Their common feature is the extensive use of vegetation to provide ecological services, such as improved drainage, lesser heat effects, bigger biodiversity and greenhouse gases sequestration.

67. See Lucas Rosse Caldas, Andrea Santos and Luan Santos. "Como Tornar as Cidades Mais Inteligentes Diante das Mudanças Climáticas e Pandemias? " 05 Abr 2020. ArchDaily Brasil. Accessed 13 May 2020. <https://www.archdaily.com.br/br/936764/como-tornar-as-cidades-mais-inteligentes-diante-das-mudancas-climaticas-e-pandemias>

However, these retrofits of the built environment are slow, large-scale operations. They are also bold initiatives that require massive economic investment and a wider net of cooperation between many different agents, practices, communities and institutions.

This, in fact, is the stuff of all those Green New Deals that have started to take over the political discussion across the Western hemisphere.

6.
All The Green New Deals

Representative Alexandra Ocasio-Cortez talking about the Green New Deal

When public discussions heated up around the 116th United States Congress proposal for a Green New Deal in early 2019, American architects were more than ready to follow suit. If still vague, the breadth of the House resolutions sponsored by instant celebrity Representative Alexandra Ocasio-Cortez with Senator Ed Markey, was too juicy and heartening to be missed by all of those involved in the mighty American construction industry.

The resolution asked for an infrastructural overhaul not seen since the aftermath of the Great Depression and President Roosevelt's muscled response to it in the form of the well-known "New Deal." While it tackled the *Great Transition* to renewable energies and full decarbonisation by 2030, the proposed Green New Deal promised trillions of dollars in new jobs, affordable housing, and many other wonders. In between plenty of equity stimuli, its measures included "building resiliency against climate change-related disasters," "repairing and upgrading the infrastructure," and, most appealingly, "upgrading all existing buildings in the United States and building new buildings to achieve maximum energy efficiency, water efficiency, safety, affordability, comfort, and durability."[68] While simultaneously addressing climate change and economic inequality, this legislation package was simply too mouth-watering to be overlooked by architects.

68. See "116th Congress H.RES.109," 7 Feb 2019. congress.gov, Accessed 19 May 2020, https://www.congress.gov/bill/116th-congress/house-resolution/109/text

Two weeks before the New Green Deal was presented to Congress, *Architectural Digest* was already asking what it would "look like for architecture," advancing that such an audacious plan would "extend well beyond the ecological."[69] In fact, it could be the case that, soon, a building wouldn't just be a building anymore. Instead it would have to be a mini-power plant. But the implications for architectural practice could be broader. Considerations that previously were apparently unlikely, such as community resilience, could become a new tenet for architectural design.

69. See Meaghan O'Neill, "What Would a "Green New Deal" Look Like for Architecture?," 22 Jan 2019. Architectural Digest. Accessed 19 May 2020, https://www.architecturaldigest.com/story/green-new-deal-architecture

70. See "AIA Supports Green New Deal Framework", 8 Feb 2019, The American Institute of Architects. Accessed 19 May 2020, https://www.aia.org/press-releases/6105450-aia-supports-green-new-deal-framework-

On the same day the resolution was introduced, the Architect's Newspaper presented a survey it conducted amid architects. Its premise was simple: since the overwhelming threat of climate change could be achieved by "only a government-led, war-like industrial and economic mass mobilization effort," how would architects get on board and benefit of such "top-to-bottom reworking of the U.S. economy?" Practitioners contributed suggestions to the resolution that ranged from tax incentives for the reuse and retrofit of existing buildings, to funding giving priority to long-term sustainability benefits. Among concerns of inclusion or social and racial equity, many highlighted the opportunities of affordable housing. The Architecture Lobby, which was soon to launch its own manifesto in favour of the Green New Deal, recalled the right to work. Others, like Vishaan Chakrabarti, insisted on the virtues of densification. The appeal of the Green New Deal did fit into many different architectural agendas.

On the day following the Congress presentation, the American Institute of Architects (AIA) was the first professional organisation in the global arena to declare their support of the measures. In a press statement issued on February 8, the AIA president William Bates applauded and encouraged the Congressional efforts. Most importantly, he repeated that "by investing in infrastructure, upgrading the existing building stock, and improving resilience in the built environment," progress could be made towards a critical carbon neutral future, as well as AIA's own 2030 Commitment goals.[70]

The whirlwind of reactions continued in the architectural media.

However, by the beginning of April, the resolution actually failed to advance in the United States Senate, and excitement started to wind down. *Architect*'s editor-in-chief, Ned Cramer, still proposed that it was time for architecture to mobilise. But before it acted, the field needed "a serious debate" and it needed to "build consensus around the best solutions for climate change and the other crises addressed by the Green New Deal."[71]

71. See Ned Cramer, "What's The Big Green Deal?," Architect Magazine, April 2019. Also accessible online with the title "Making Sense of the Green New Deal," 17 April 2019. architectmagazine.com. Accessed 19 May 2020. https://www.architectmagazine.com/design/editorial/making-sense-of-the-green-new-deal_o

Basically, as it had happened to legislators, architects too were asked to go back to the drafting table. Nevertheless, even if the political game fizzled out, it took only a few months for new reflections to emerge in architectural academia.

A conference held at the University of Pennsylvania in September was quick to address the topic of "Designing the Green New Deal." Billy Fleming, the director of the Ian L. McHarg Center at the Stuart Weitzman School of Design, and one of the event's organisers, focused on how legislation could enable the large infrastructure projects central to the proposed reform. The conference's guest keynote, however, was more adamant.

As reported on *Curbed*, Kate Orff, the renowned landscape architect and McArthur fellow, was clear on how the Green New Deal was really about "designing an entirely new world." What she deemed the power of imagining "speculative infrastructure" was being called upon to influence policy and generate the interdisciplinary coalitions required to achieve the goals of the Green New Deal. But this would also require newly minted government agencies to open the way to restoration and regenerative practices that would address "environmental racism."

Indeed, if there was a benefit that the Green New Deal discussions brought to the design field, it was certainly their acceleration in linking emerging environmental concerns to other social and economic issues entering architecture discourse. The debate was no longer only about making architecture "greener," as it had been the case for decades. It was also about a broader awareness of the delicate balances taking place between damaged ecosystems and vulnerable socio-economic structures.

Resilient Boston Harbor
Vision by SCAPE

As author Phillip Tabb portrayed in a critical account of a long history of *Greening Architecture*, along the second half of the 20th century this discipline was long engaged with the need to accommodate better environmental practices. But, to the exception of those radical deviations we have already located at the turn to the 1970s, this did not necessarily engage a holistic ecological vision such as the one that slowly moulded on the pressures of a climate emergency combined with growing global inequity.

As Tabb concurs at the beginning of his comprehensive narrative, "green architecture" mainly started as a practice that advanced "from rationalist, performance-based and remediating measures in response to particular unsustainable concerns." Along the way, it contributed to some formal or technical incorporations into architecture movements. Tabb indeed defends it "served to explore environmentally oriented canons with new tectonic and emergent architectural languages."[72]

72. See Phillip James Tabb and A. Senem Deviren. *The Greening of Architecture : A Critical History and Survey of Contemporary Sustainable Architecture and Urban Design*, Ashgate, Surrey: 2013.

Yet, to the fleeting exception of the avant-garde deviations we have already mentioned, this never meant that the greening effect took the lead in a determinant architectural narrative — which this author would like to see in it. If the greening was ever "pervasive," I would counter, this was due to the ways in which the easy motto of sustainability took hold of mainstream architecture's more commercial or greenwashing discourses. As Tabb recognises, though, it was only much more recently that what passes by green architecture evolved into "far more encompassing ecological and systemic processes cutting deep across contemporary culture."

In such contemporary situation, the Green New Deal appeared as a contextualising political framework that helped position impulses of greener architecture within a broader social ambition. From worn down claims of sustainability we aimed to move to notions of resilience or regeneration. Further, the mirage of a Green New Deal offered hope that such greening impulses would be supported by an economic policy that would correct market failures and create favourable conditions for an investment in more radical forms of innovation and conciliation.

While trillions would be made available, the effective implementation of a Green New Deal would produce a major economic and social upheaval. In such *Great Transition*, technical and formal innovations would be essential for architecture to help operate the scale of transformations required for full decarbonisation in a relatively short time. But conciliation of the opposing forces of market-driven or ecological imperatives would also be crucial, so that design practices are able to help correct social inequities along the way.

Following this logic to its origin, it is therefore important to note that the Green New Deal was not an invention of democratic politicians in the context of the American bipartisan politics circa 2018. The idea in fact gained traction a decade earlier, around the time of the previous global financial crisis, i.e. the Great Recession starting in 2007/2008. And, in fact, it was also around that period that the notion was first brought to impel architectural discourse.

Today, Green Parties claim they were the first to call for a Green New Deal, with an initial task-force supposedly created for that purpose in 2006. Yet, it is New York Times journalist Thomas Friedman that is credited with first bringing the idea to mainstream discourse. And already in 2007, the language came close to what we hear today.

Friedman kicked off with the growing effects of climate change, rather than the financial crisis. He strictly linked the need of a green version of the New Deal to a massive economic and political effort if we indeed were to transition to renewable energies. This would require more than your individual rooftop solar panel. As he wrote at the time, it required "a huge industrial project — much bigger than anyone has told you," if we really wanted to "green the world" and, simultaneously, "spur our economy into the 21st century."[73]

This call has certainly triggered some more thinking. That same year, the United Kingdom saw the influential Green New Deal Group coming together and publishing their first Deal proposal in July 2008, only a couple of months before the subprime mortgage crisis exploded globally. And they did see it coming.

As the opening statement of *A Green New Deal* goes, the global economy was facing "a triple crunch." This arose from "the combination of a credit-fuelled financial crisis, accelerating climate change and soaring energy prices underpinned by an encroaching peak in oil production."[74] And the resulting "perfect storm" could be expected to match the impact of the Great Depression of 1929. Then — as hopefully now under the added crunch of a global epidemic and the sudden suspension of consumption as the world's primary economic drive — the massive endeavour of a Green New Deal was seen as the only possible way out.

The ideas contained in that early proposal took more than a decade to make its way into the UK Parliament, where only shortly after the US Congress resolution a bill was also introduced in March 26, 2019. Under the designation of *Decarbonisation and*

73. See Thomas Friedman. "A Warning From the Garden", 19 January 19, 2007, *The New York Times*. https://www.nytimes.com/2007/01/19/opinion/19friedman.html

74. See Green New Deal Group, *A Green New Deal*, New Economics Foundation, London: 2008. Accessed 20 May 2020, https://greennewdealgroup.org/wp-content/uploads/2019/06/a-green-new-deal.pdf

Economic Strategy Bill, the legislative proposal set out "the duties to decarbonise economy and eradicate inequality."[75]

75. See *Decarbonisation and Economic Strategy Bill,* 26 March 2019. United Kingdom Parliament. Accessed 20 May 2020, https://publications.parliament.uk/pa/bills/cbill/2017-2019/0365/190365.pdf

Meanwhile, however, the debate had heated up between economists, political commentators, ecologists and international policy makers. By April 2010, an article in *Nature* magazine checked how was "the Global Green New Deal going," and it concluded that, as usual, things were harder to implement than they seemed on paper. Under that "global" ambition, an initiative had been launched by the United Nations Environment Programme with the intent to "stimulate economic recovery and create jobs, enhance the livelihoods of the world's poor and lessen carbon dependency and environmental degradation." G20 countries embraced the initiative, and China and South Korea dedicated significant stimuli packages to "green projects." But with the world facing multiple crisis, the recommended "expenditure of 1% global gross domestic product (GDP) on green initiatives" never really took off for most.

As economist Edward Barbier stated, even when argued that every $1 billion invested in energy efficiency and clean energy could generate savings of up to $450 billion, a lack of "supportive policies to ensure the long-term economic viability of sustainable practices" would inevitably drive the programme to fail. But, according to the author of both plan and the article in *Nature*, an even worse failure of the G20 in implementing the UN's Global Green Deal was its neglect to "address the economic and environmental vulnerability of the world's poor."[76]

76. See Edward Barbier, "How Is the Global Green New Deal Going?," Nature, Vol. 464, Iss. 7290, London: 2010.

In this, the good intentions of all the Green New Deals risked looking like just another form of rhetorical greenwashing. This is why, when the author of this article revisited the recent rebirth of the Green New Deal, he suggested that this time around states would need to "finance the new policy sustainably."[77] After slashing fossil fuels subsidies, a green and equitable transition would only last if the money kept flowing — and rather than by imposing deficits on future generations, money could keep flowing if carbon taxes became widespread.

77. See Edward Barbier, "How to Make the Next Green New Deal Work" Nature, Volume 565, Iss. 7737, London: 2019.

In 2008, however, the theory of a Green New Deal continued to make sense and generate hope — and the architectural world did take notice. Young architecture critic Owen Hatherley, then writing at *Building Design* magazine, suggested that the Green New Deal could help tackle the economic depression that was quickly falling upon both the building industry and architects.

While also recalling the unexpected benefits of the 1929 crash on the sobering of architectural styles, Hatherley stressed how the public spending of Roosevelt's New Deal "went on public works, not handouts to banks." The implicit hope would thus be that contemporary calls for its new green version would also bring "social housing and infrastructure on an unprecedented scale." While these measures once "averted disaster," now they could keep architects busy with "a less destructive and wasteful economy."[78]

Still, a less economy-driven, more ecologically prone view of the New Green Deal was also emerging in practice itself. In the work of Barcelona-based architect Enric Ruiz Geli, in particular, the Green New Deal served as the incitement for a more radical investigation.

The *Green New Deal* architecture experiment started in 2009, as an exhibition at the then revamped Arts Santa Monica, in Barcelona. The project symbolically prompted the institution to make a political shift and, in the words of its new director, open the space for citizens by "linking the fields of science and technology to projects of radically humanistic and innovative creation." As the head of the arts centre understood it, "culture itself" should lead the change "in the social, economic and cultural models regarding sustainability, the use of energy, the commitment to fight global warming, and the obligatory challenge of reducing CO2 emissions."[79]

For the institution's re-opening exhibition, Ruiz Geli proposed a technical analysis of the arts building itself. Data on internal heat flows, insulation, as well as contextual conditions, were mined and recorded as a pathway to investigate what

78. See Owen Hatherley, "A Green New Deal Can Tackle Depression" Building Design, October. Tonbridge: 2008.

79. See Vicenç Altaió, "Presentation," in Enric Ruiz Geli, (ed.), *A Green New Deal: From Geopolitics to Biosphere Politics*, ACTAR Publishers, Barcelona: 2011.

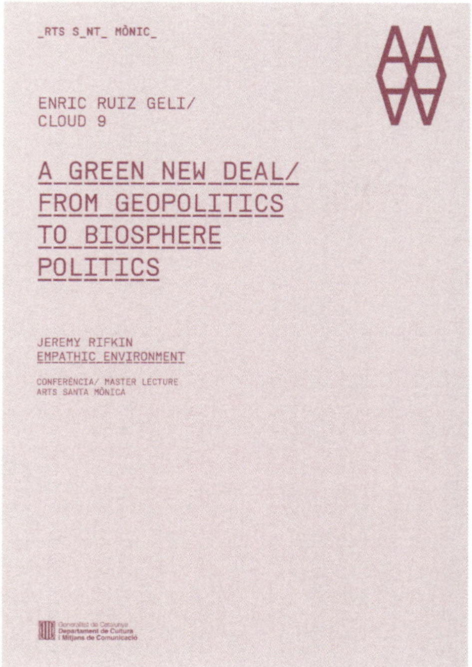

A Green New Deal: From Geopolitics to Biosphere Politics by Enric Ruiz Geli / Cloud 9

technologies could allow the building to reduce its carbon dioxide emissions. Halfway between scientific experiment and art installation, the building was then enveloped by hundreds of suspended plants, sensors and projected images of its future use as centre for knowledge and energy. By bringing the ephemeral intervention to the institution's facade, the architect aimed at reaching its audience directly at Barcelona's busy Ramblas. The project unfurled as a manifesto calling not only for a wider awareness, but for people's direct involvement in combating the climate crisis.

In the book that followed, Ruiz Geli reminded us that "the number one cause of global warming is Architecture." Thus, architects too must assume their responsibilities and "move from geopolitics towards biosphere politics."[80] This explained the intervention in the Santa Monica arts centre: with the addition of the living plants, as well as a deeper understanding of the building's organic and mechanical interference with the surrounding atmosphere, a new reality was produced. No longer strictly based on the formal and physical characteristics of 'place,' architecture sought to connect with the more complex politics of its enveloping 'biosphere.'

Aiming at off-the-grid solar and wind energy production, the building should now learn from natural processes. Recurring to sensors registering temperature, humidity, and the light's energy, the new green skin added to the arts centre was intended to offset the building's permanent CO_2 production. In the exchange with architecture's mineral nature emerged the metaphorical "green new deal" that gives title to the intervention.

As opposed to the logic of the hi-tech architecture of the 80s and 90s, this "green new deal" represented a sort of lo-tech strategy to retrofit existing structures with a new ecological prowess.

80. See Enric Ruiz Geli, "Towards an Empathic Architecture," in Enric Ruiz Geli (ed.), *A Green New Deal: From Geopolitics to Biosphere Politics*, ACTAR Publishers, Barcelona: 2011.

At its height, hi-tech architecture had fetishized structural and mechanical elements as a formal and expressive device, prioritising the spectacle of advanced technological systems. As Philip Tabb tells us in *The Greening of Architecture*, this trend eventually caught up with environmental systems that provided a much-needed sense of sustainability. By way of its technological fixtures, "green should become standard practice." Yet, these were static processes mainly aimed at efficiency and durability. Only with the emergence of what Tabb calls 'eco-technology' did this techno-romanticism turned to dynamic systems that echoed "the changeable nature of onsite resources, diurnal cycles and inclement conditions."

It was within this tendency that Ruiz Geli's experiment was to be understood. Indeed, as Tabb would put it, this architect's approach "harnessed and subtly manipulated natural forces to leverage their beneficial environmental effects and expressed them explicitly." However, his "green new deal" project had also a DYI quality that appealed to more than just integrating and showing off available engineering add-ons. Many of the technologies deployed were created in-house or customised by Ruiz Geli's practice Cloud 9. They spoke of an inventive ingenuity that echoed the realignments required at every kind of scale and on every front by a wider Green New Deal. As opposed to the high-tech tradition, its low-cost, crafts strategy aligned more properly with a massive economic upheaval in which every person and every community would be asked to contribute with whatever resources they had at hand. 'Sustainability' did not have to come with a scandalous upfront cost.

It was not per chance that the project involved Jeremy Rifkin, one of the economists who then led the public discussion on the urgency of a Green New Deal for a potential "third industrial revolution." Speaking to the press at the exhibition's opening, Rifkin lauded the arts centre for the transformation of its headquarters into a living laboratory of such revolution. Usually thought of an institution to preserve the past, the museum was here instead to be converted into a power plant and a oxygen production unit — and to become, through the combination of art and science, "a frame of reference in an ongoing journey to explore the human journey." For Rifkin, this museum experiment was an imaginary exploration of "our empathic responsibilities to each other, our fellow creatures and the larger environment that we are in." Ultimately, the experience would serve to "get us to biosphere consciousness."[81]

To this cultural shift corresponded, in the field of architecture, an overcoming of the 'sustainable architecture' paradigm. As the author of *Greening Architecture* has it, that prior paradigm was focused "on the immutable formal and material characteristics of a building's design and associated technologies," with every system directed at "a singular purpose." In the new paradigm of "ecological architecture," on the contrary, invisible flows addressing dynamic ecosystem processes were to become architecture's core conceptual drive. And with those ecosystem processes including the forces that mould human social networks, a revealing correspondence was established to the overall economic and political aspirations of all Green New Deals proposals.

In a more formal address, Rifkin went on to suggest that, with the prolonged sunset of the previous "great energy regime" — the twilight of what future generations would call the "Fossil Fuel Age"— we faced four major crises: climate change, global debt and inflation, geopolitical instability and the looming of peak oil production itself. And the enormity of those crises required a major "economic game plan," which would only be feasible within the context of a third industrial revolution centred on distributed energy production.[82]

Rifkin opposed the notion of distributed energies to prevailing "elite energies," which typically require huge political and military investments, massive capital flows, and the inevitable top-down management conducted between nation-states and transnational corporations. For a true revolution to happen, and for a Green New Deal to actually succeed, such *status quo* would

81. See Jeremy Rifkin, "Press Conference at Arts Santa Mónica," in Enric Ruiz Geli, (ed.), *A Green New Deal: From Geopolitics to Biosphere Politics*, ACTAR Publishers, Barcelona: 2011.

82. See Jeremy Rifkin, "Empathic Environment," in Enric Ruiz Geli, (ed.), *A Green New Deal: From Geopolitics to Biosphere Politics*, ACTAR Publishers, Barcelona: 2011.

have to be overcome and energy would have to learn from the way the Internet today operates, with its distributed nodes and multiple redundancies.

Following on Rifkin's ideas, at the heart of the Green New Deal would then be the ways in which architecture, as policy-making or any other economic sector, could contribute to the transition to a new economic model based on ecologically driven distributed energy systems. This should not correspond to applying the same old top-down approach to renewable energies — as, in fact, it is already happening — but rather to the adoption of a true, locally based energy commons. There would still be back-ups in place through regional or national grids, but an essential part of the energy needs could be supported by environmentally sensitive local management, at the level of communities and municipalities. In such scenario, as we have already hinted at, the same buildings that are today weapons of ecocide could now be nodes for such an energy distribution network: as well as biofactories and CO2 drawdown machines, buildings would be the power plants for the Green New Deal.

In a plan developed for the European Union's *Great Transition*, Rifkin precisely proposed that the 191 million buildings existing in that territory would provide for a decentralised, collaborative energy strategy. And even if such a massive retrofit operation would not be entirely managed by architects, architecture practice could surely lead its most innovative and model-setting aspects — if indeed this practice would undergo its own adaptive repurpose. If architects would embrace such change, there would be both an economic and a creative opportunity for a radical shift.

As Rifkin advanced, "the coming together of the Internet and renewable energy — the democratisation of both information and power in one matrix — is what changes the whole frame of reference for everything we do in society."[83] And because of the centrality of buildings to such transition, in the Green New Deal move to a post-carbon built environment there is a potential for

83. See Terry Waghorn, "Jeremy Rifkin's Third Industrial Revolution," 12 Dec 2011. Forbes. Accessed 24 May 2020, https://www.forbes.com/sites/terrywaghorn/2011/12/12/jeremy-rifkins-third-industrial-revolution/#3a5c206546f1

the conceptual overhaul of architectural language at the level of the one carried out by early 20th century modernists.

Much as early modernism responded to contextual changes made available by the First Industrial Revolution, the advancement of a Green New Deal would create the political and economic opportunity for architecture practice to get on board the Third Industrial Revolution. But this would implicate a greater integration with change happening around us — including the move to incorporate a deeper ecological response in each single new or retrofitted building.

As the other editor of *Greening Architecture* suggests, "the architects' responsibility of thinking in context and networks goes way beyond solving the problem of isolated "stand-alone" structures." Evolving from earlier forms of eco-architecture that were unable to create "living relations with the surrounding ecology," A. Senem Deviren writes, "new landscapes are inclusive and require network thinking."[84] Thus, buildings will be made "a part of the living environment" by way of new interdisciplinary collaborations, but also, I would add, by the ability of today's architects to fully absorb the cultural transformations implied by the *Great Transition* — as much as their predecessors one century ago took on similar inputs to trigger the Modern Movement.

84. See A. Senem Deviren, The Global Landscape of Green Architecture, in Tabb, Phillip James, and A. Senem Deviren. *The Greening of Architecture : A Critical History and Survey of Contemporary Sustainable Architecture and Urban Design*, Ashgate, Surrey: 2013.

Absorbing cultural transformations, nevertheless, does not mean only to be able to translate some of their aspects into new technical impulses or a new formal language. It also implies addressing and incorporating novel eco-social concerns, ideals of environmental and social justice that all the Green New Deal proposals carry with them.

As such, in his recent "Preliminary Notes on a Design Politics for a Green New Deal," sociologist Damian White investigates how "the emerging field of design for transition" should be encountering those movements that, within the scope of the late Green New Deals, fight for the just dimensions of that transition. As he opens it, "post-carbon futures are not simply going to emerge through protest and policy shifts alone."

Even a "just transition" will also "have to be imagined and built, fabricated and realized, coded and created."[85] And this requires not only abundant creative impulses, but also practices that welcome different dialogues, new political responsibility and effective conceptual adjustments.

First of all, White suggests, design for transition has to overcome the small level of engagement it still commands. This includes expanding the conversations with "design activists, transition town advocates, radical municipalists and commoners." It includes reinforcing intellectual and political alliances with environmental justice movements that question social and economic inequities, as much as the extractive, patriarchal nature of Western colonisation.

Besides the usual dichotomy between the "breakthrough" of a techno-fix to decarbonise business as usual, or the "breakdown" of "survivalist communism;" besides the basic opposition between "acceleration" of *ecomodernists,* and the "deceleration" of *degrowthers*, White proposes, one should previously accept that, in the spirit of an overreaching Green New Deal, "all conceivable programs for just transition are going to be socio-technical in nature, multi-scalar and, by definition, concerned with designing low-carbon futures."

Design options for a just transition and the promised context of a Green New Deal would necessarily be hybrid and keen to open new pathways that combine "state-led action *and* grassroots innovation." Much as in the notion of tactical urbanisms and its addressing of growing urban inequality,[86] design must find a middle ground between top-down strategies and bottom-up initiatives, so as to counter the disconnect that hides in the ability to decarbonise while, *at the same time*, keeping the *status quo* intact. Be it in the norms regulating the built environment, be it amid the socio-economic forces driving modern societies around incessant growth and consumption, architecture and city planning would have to find ways to achieve new balances between top-down and bottom-up, between hi-tech and lo-tech, between the market and social equity.

85. See Damian White, "Just Transitions/Design for Transitions: Preliminary Notes on a Design Politics for a Green New Deal." in Capitalism Nature Socialism, Volume 31. Routledge. London: 2020.

86. See Pedro Gadanho, (ed.) *Uneven Growth: Tactical Urbanisms for Expanding Megacities*, MoMA, New York: 2014.

While mostly offering an historic survey of the *just transition* concept, including its appearance in the design counterculture of the 1970s, White finally suggests that a design politics for such a scenario "must explicitly push back against neoliberal design and managerial and corporate visions of 'design thinking'." The RISD scholar offers examples such as that of Ezio Manzini and the DESIS lab network, the Carnegie Mellon's Transition Design research group, or the Decolonising Design group. With the contemporary work of those practices, he illustrates how communitarian design innovations, participatory processes, new digital tools and scenario thinking come together to reveal "cosmopolitan-local transition paths for low-carbon futures." While partially speculative, the interventions he describes resulted in practical, local, but scalable projects, which typically included neighbourhood gardens and care centres, maker-spaces, fab-labs or co-operative housing.

For sure, over the past years these experiments have offered a glimpse of potential pathways for nurturing interactions between designers and specific communities. Rather than the vehicle of top-down impositions, designers may be seen here as mediators that abilitate a community to create its own defence mechanisms against the disruptions of a deep ecological crisis. Yet, even if such operations can be replicated in many similar situations, they still seem like only a small piece of the environmental design puzzle our society is faced with, as it enters the *Great Transition*, or even as it seeks to benefit from the massive financial investment of a potential Green New Deal.

Indeed, White is on target when he finally asks how we can "design new public institutions and new forms of public ownership to expand the range of participatory design directed toward urban and infrastructure transition." Beyond formal and technical adaptations, this an essential conceptual conversion for an architecture practice that reflects the multiple needs of a massive decarbonization operation. In that case, all those prospective Green New Deals may provide more than just the right economic support framework to accommodate a radical reorientation of architecture's business models. They may also provide a necessary impulse for the dramatic creative shifts that are necessary when it comes to contribute towards a more equitable, distributed and ecologically sound energy transition.

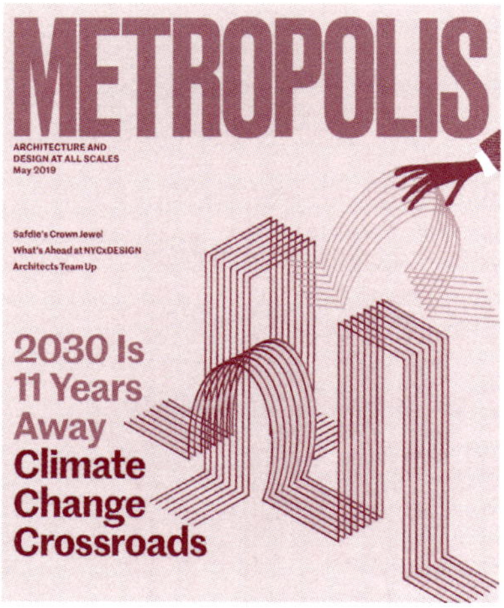

Metropolis Magazine (May 2019)

As another writer rightly put it in a Metropolis magazine essay, "for the moment, the Green New Deal exists only as a hazy resolution." However, its importance could lie "in its ability to galvanise activists and designers alike." Exploring the evolution from notions of resilience to those of regeneration, Stephen Zack posits how "the Green New Deal is open to further articulation."[87] In its recent versions, for instance, the Green New Deal proposals mention "regeneration only in relation to agricultural practices" — but translating such indications to the realms of architecture, landscape design and urban planning can prove to be inspiring.

Also, where the sustainability-related certifications of yesterday were still mainly concerned with the ability to maintain a given state of affairs, a.k.a being resilient, new building standards could already be aiming at delivering "regenerative systems as intergenerational equity." Emerging certification systems such as the SITES initiative or the RELi standard now include regeneration as one of their central guiding principles. Drawing on recent urban-ecological studies — and reminding us of the broader environmentalism of the built landscape described in the previous chapter — they propose "regenerative frameworks," which not only admit properties of "biodiversity, ecosystem feedback, redundancy, change, and ephemerality," but also seek "to enhance" these deep structural elements of a given territorial unit.

If some of these certifications systems were first intended to provide guidelines for the field of architecture landscape, they represent a lesson in designing "with" nature, instead of "against it," which deserves urgent transcription to the practices of building design. As Zacks points out, they represent an ambition to

87. See See Stephen Zacks, "Long Term Plans: To Build for Resilience, We'll Need to Design With—Not Against—Nature," 8 Jan 2020. Metropolis, Accessed 19 May 2020 https://www.metropolismag.com/architecture/landscape/resilience-sustainability/

integrate a "broader enterprise of reversing ecological degradation." But as much as they highlight "the centrality of regulation in achieving this aim," they again recall the importance of having the framework of a Green New Deal in place, at least as a galvanising force that fosters the possibility to start an important change in architecture practice. Because where "new construction would be regulated according to ecological mandates," there would also be a strong economic, legal and social incentive to start such change.

In this context, however, where government leadership may still lack to effectively implement all those Green New Deals, it is comforting to realise that cities themselves have been enthusiastic to shore up their own roadmaps to achieve similar goals.

In October 2019, titles such as "Big City Mayors support Global Green New Deal" made the headlines. This came along with dire warnings that the planet was already "in the grip of a climate crisis." Thus, prompted by the C40 Cities Climate Leadership Group, almost 100 cities representing "one-twelfth of the world's population and one-quarter of the global economy" announced in Copenhagen their priority to "protect the safety of their citizens."[88]

88. See "Big City Mayors Support Global Green New Deal," Environment News Service, Lincoln: 14 Oct 2019.

Referring to the failure of world leaders to follow up on the Paris Agreement, — and thus trigger the actions that the United Nations Framework Convention on Climate Change (UNFCCC) made explicit — it was precisely the Paris Mayor, Anne Hidalgo, who urged that "all decision-makers must take responsibility for making [the Global Green New Deal] a reality." As the report went, this would comprise "inclusive climate action at the center of all urban decision-making to secure a just transition." It would also demand "correcting long-running environmental injustices for those disproportionately impacted by the climate crisis." On the other hand, London Mayor Sadiq Khan added what was hitherto well-known: cities were already leading the way to a semblance of a Green New Deal with their considerable investments in the green economy, decarbonization and sustainable jobs.

With their authority over building codes and permits, cities had always been more prone to exercise an influence on how architecture practice would adapt to different circumstances — and this would be no different with the challenges presented by climate disruptions and ecological disasters. City authorities have typically been the first to acknowledge the dangers and inequities caused to their inhabitants by such challenges, and their mandate of proximity determines that they have to respond in much more immediate ways. Recalling that cities had "substantial jurisdiction over carbon-emitting buildings," Kate Harrison, a council member for Berkeley, California, joined the debate a few days later, urging that, although coordination between national and city levels was mandatory, cities could be more agile in implementing the extensive measures implied by a Green New Deal.[89]

89. See Harrison, Kate, "Cities can lead the fight for a Green New Deal." Daily Californian, 15 Oct 2019. University of California, Berkeley.

As a result of these insistent calls, at the turn to the 2020s, the pleads for the urgency of a Green New Deal extended even to sectors that would normally not embark in such socially inclined adventures. Even for corporations and conservative economic outlets, the subject acquired a respectable weight and elicited serious debate. Eventually this wave also involved the critics that represented architectural discourse in the most circumspect newspapers.

As architecture critic Edwin Heathcote put it the *Financial Times* in late 2019, in a context of massive change the Green New Deal hence remained a credible "attempt at reframing green imperatives in a Rooseveltian image, as a boost to the economy rather than a brake."[90] In his analysis, Heathcote linked that reform with the imperative to adopt the circular economy in the building sector. He referred to a successful housing retrofit scheme by French architects Lacaton & Vassal and Frédéric Druot. The Tour Bois-le-Prêtre, in Paris, represented a good example of "moving beyond a 'take-make-waste' model of consumption," providing for both the conservation of embodied carbon and an enlightened take on affordable housing. And yet, it must be noted that this project was also influential because its authors and architectural aesthetics rang a bell within the inner circles of the architecture discipline.

90. See Edwin Heathcote, "There's nothing new about tomorrow's city", 5 Dec 2019, Financial Times. Accessed 19 May 2020, https://www.ft.com/content/6c60ec0a-f9b1-11e9-a354-36acbbb0d9b6

Tour Bois-le-Prêtre by Druot, Lacaton & Vassal

In fact, that particular design and its successful diffusion proved that, when examples come from the most legitimate players in the field, they can become truly determinant in more practices adhering to other imperatives. As such, it was particularly relevant that Lacaton&Vassal became the recipients of the Pritzker Prize in 2021.

But Heathcote concluded with a different idea: besides now being power utilities, as mentioned earlier, buildings might also soon start being taken as banks for materials to be re-used. While for now the U.S.-based New Green Deal prevailed only as the stuff of presidential debates — and the European Union own package was suddenly buried under the economic recovery imposed by the Covid19 outbreak — other smaller initiatives could now be the carriers of the efficient retrofit flame.

The EU-sponsored *Buildings as Material Banks* initiative, which should really be affectively known as Bambi, was the case at hand. As in programs targeting species in risk of extinction, the initiative investigated the possibility of embedding identification chips in building components, thus allowing for their reuse in new constructions. Hence, demolition could become part of an efficient recycling process directed at the built environment.

If this digital tracking mechanism for the circular economy is still a bit of a mirage, the fact is that some architectural practices over the past decade have already turned to the recycling of industrial materials into buildings as a possible path to counter the wasteful "planned obsolescence" of a consumption-driven economy. Following on the footsteps of pioneers Mike Reynolds and Rural Studio, Rotterdam-based practice Superuse Studios devised a collaborative mapping system to identify regional overflows of waste materials that could be employed in new architectural projects. In the same vein, Heathcote added the examples of Brussels-based Rotor Deconstruction or the Aarhus School of Architecture's "Bank of Materials."

The Miele Space Station by Superuse Studios

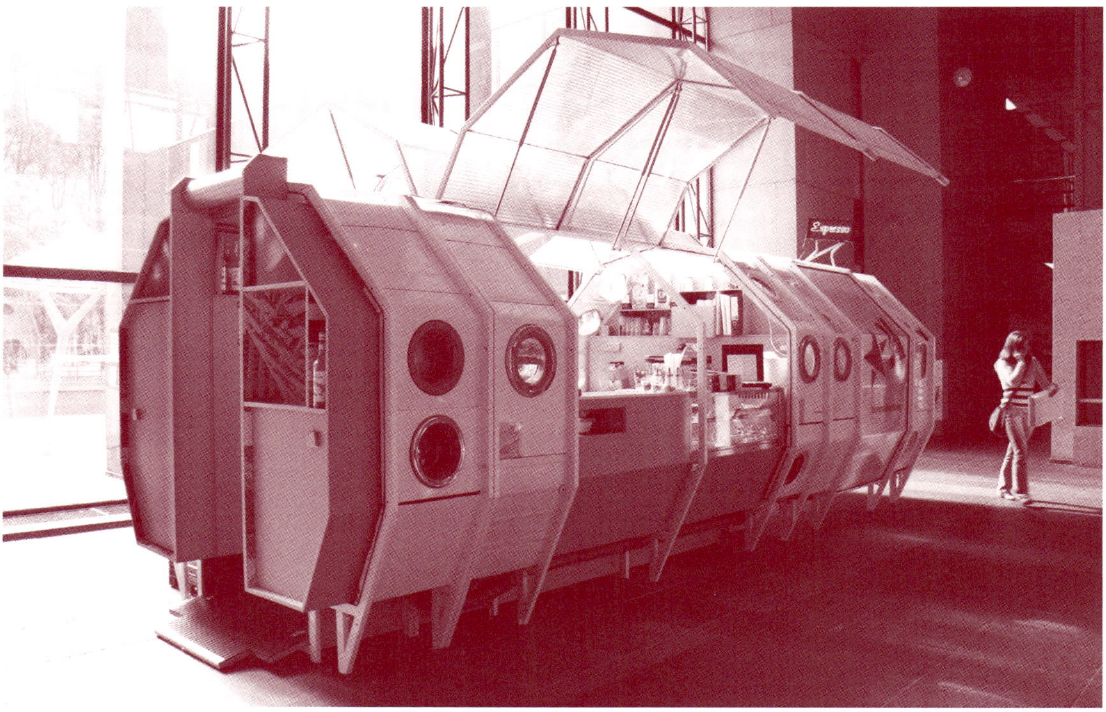

These, after all, were but a few illustrations of one potential path to fully operationalise a Green New Deal — as soon as that resolution gains democratic appeal and political support enough to become a reality.

Once enacted, a Green New Deal would surely stimulate other innovative avenues through which architecture practice could contribute to a massive collective effort. Seeds of invention may be hiding in creative retrofitting or adaptive reuse. But these are only a fraction of the possibilities that should arise from a reorientation of practice to the ecological and social dictates of an effective Green New Deal. And this may be especially galvanising for a generation that, while recently entering the profession, are both the most vulnerable to the coming disruptions of a climate crisis, but also the enablers of a desirable just transition into a new paradigm.

This is the reason why the reaction to recent Green New Deal announcements by a young activist group such as The Architecture Lobby (T-A-L) becomes particularly interesting. Going through the US Congress Resolution 109 in great detail, The Architecture Lobby effectively expanded on the legislative initiative bringing it to bear more radical implications for the architecture profession.

In their June 2019 statement on the Green New deal, they swiftly open with a demand for a reform in architectural practice: "in order to tackle decarbonization efforts more effectively, the way we work and the way the profession is structured must change." Springing from a collective of precarious workers in the architecture field, the sweeping tone immediately discloses how, even in the design arena, the Green New Deal fits an agenda of social and labor reform.

T-A-L's adherence to the spirit of the legislative initiative goes further than most, asking for "a code of ethics" to support its "social, economic, and environmental aspirations." This would include the needs to "reject the current model of practice as a service profession responding primarily to private capital," or to "redefine sustainability to acknowledge the economic, social, racial, and class-based dimensions of the climate crisis."[91]

91. See "T-A-L Statement on the Green New Deal," June 2019. The Architecture Lobby. Accessed 19 May 2020, http://architecture-lobby.org/project/t-a-l-statement-on-the-green-new-deal/

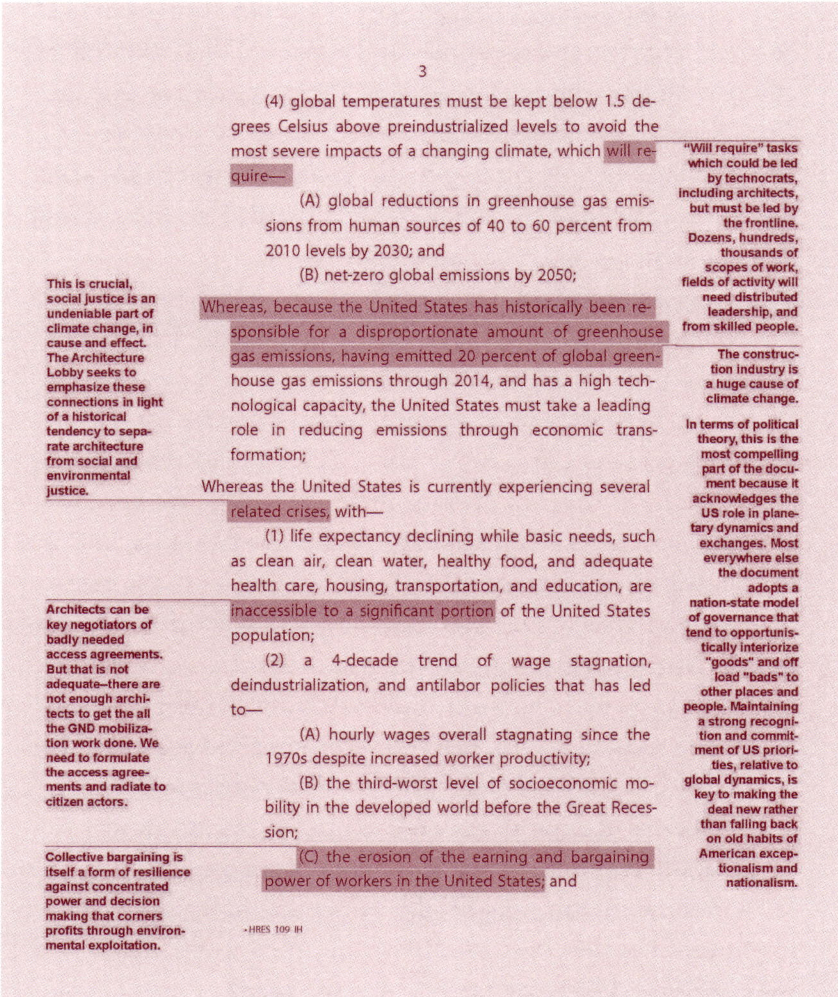

3

(4) global temperatures must be kept below 1.5 degrees Celsius above preindustrialized levels to avoid the most severe impacts of a changing climate, which will require—

(A) global reductions in greenhouse gas emissions from human sources of 40 to 60 percent from 2010 levels by 2030; and

(B) net-zero global emissions by 2050;

Whereas, because the United States has historically been responsible for a disproportionate amount of greenhouse gas emissions, having emitted 20 percent of global greenhouse gas emissions through 2014, and has a high technological capacity, the United States must take a leading role in reducing emissions through economic transformation;

Whereas the United States is currently experiencing several related crises, with—

(1) life expectancy declining while basic needs, such as clean air, clean water, healthy food, and adequate health care, housing, transportation, and education, are inaccessible to a significant portion of the United States population;

(2) a 4-decade trend of wage stagnation, deindustrialization, and antilabor policies that has led to—

(A) hourly wages overall stagnating since the 1970s despite increased worker productivity;

(B) the third-worst level of socioeconomic mobility in the developed world before the Great Recession;

(C) the erosion of the earning and bargaining power of workers in the United States; and

•HRES 109 IH

"Will require" tasks which could be led by technocrats, including architects, but must be led by the frontline. Dozens, hundreds, thousands of scopes of work, fields of activity will need distributed leadership, and from skilled people.

This is crucial, social justice is an undeniable part of climate change, in cause and effect. The Architecture Lobby seeks to emphasize these connections in light of a historical tendency to separate architecture from social and environmental justice.

The construction industry is a huge cause of climate change.

In terms of political theory, this is the most compelling part of the document because it acknowledges the US role in planetary dynamics and exchanges. Most everywhere else the document adopts a nation-state model of governance that tend to opportunistically interiorize "goods" and off load "bads" to other places and people. Maintaining a strong recognition and commitment of US priorities, relative to global dynamics, is key to making the deal new rather than falling back on old habits of American exceptionalism and nationalism.

Architects can be key negotiators of badly needed access agreements. But that is not adequate–there are not enough architects to get the all the GND mobilization work done. We need to formulate the access agreements and radiate to citizen actors.

Collective bargaining is itself a form of resilience against concentrated power and decision making that corners profits through environmental exploitation.

"The Architecture Lobby, Green New Deal Working Group, 2019"

Considering that design is only one possible form of agency, T-A-L tendered that "activism and refusal" were also meaningful architectural acts. These would be some of the alternative means to "steer architectural practice toward projects and programs that promote democracy and equity in society." Asking for a critical take on "market-driven development" or architecture being "a conduit for accumulating wealth at the top," they pushed architects to "advocate for public infrastructure and just land-use practices."

Echoing the growing protests of a much younger crop of environmentalists emerging around the globe, T-A-L revealed an effective affinity with the Green New Deal implications that the *Great Transition* had, per force, to also provide the opportunity for a *just transition*. According to the Lobby, changes in physical infrastructure should be guided by much-needed transformations of the underlying social infrastructure.

Yet, curiously, T-A-L's statement also reverberated with the faint echoes of a much less discussed repercussions of the Green New Deal in architecture culture.

By suggesting that "architects cannot rely exclusively on technological solutions," they actually hinted at a core problem of the original political project. While they focused on the need to dismantle technology's apparent "neutrality," and its imbeddedness in "complex power structures," they also recalled that "the carbon footprint of technological fixes should be factored into sustainable design." They rightly envisioned that, with the climate crisis being a transversal societal challenge, "advanced climate mitigation and adaptation technologies should be open source and accessible to all." They clearly understood that technological fixes had hidden environmental and social costs, continuing to produce labor exploitation, resource extraction, and rampant economic inequality.

Inadvertently, T-A-L's take on the technological implications of the Green New Deal reminded us that there were some bigger red flags out there. Given the breath, the collective commitment and the massive investments it required, the *Great Transition* would surely entail many unforeseen obstacles. More importantly, however, by simply believing in a virtuous economic miracle supported by technological silver bullets, all those Green New Deals could be riddled with a lot of magic thinking.

7.
Magical Thinking & Silver Bullets

As brightly announced in their website's unstoppable newsfeed, multinational architecture company Foster + Partners launched its "Sustainability Manifesto" on December 6th, 2019.

The firm was as eager as always to be at the forefront of the profession's embrace of progressive innovation and technological advancement, this time in favour of a more environmentally-friendly output. They were adding to their recent pledge of carbon neutral offices by 2030 and other timely, if slightly troubled commitments.[92] The practice founded by Sir Norman Foster in 1967, now a 1,300-staff strong 'global studio,' went public with a new "methodology for greener projects." Further, the document demonstrated its "ongoing commitment towards achieving the goals of the Paris Agreement."[93]

As it turns out, the "manifesto" was seemingly more of a PowerPoint delivery, expressly prepared for a number of presentations at the UN Climate Change Conference COP25, in Madrid. For those who might not remember it, this was another of those annual negotiation marathons in which 27,000 delegates from across the globe made very little progress in establishing the operational targets of the Paris Agreement. Nonetheless, over those early December days, at the invitation of the United Nations Framework Convention on Climate Change (UNFCCC), the European Commission and others, Foster + Partners repeatedly delivered its "unique system" intended to "effectively determine carbon emissions" at every stage of a given architectural project.

Over the previous months, the firm had swiftly and smartly revamped its earlier neutrality oaths so as to address "the measurement of embodied and operational carbon" in each of their future projects. The studio claimed that the process they had collaboratively developed over the past five years now offered "greater insight into the different aspects that affect the environmental impact of a project," while also defining "tools to mitigate and lower their carbon content."

As the company's head of sustainability, Chris Tott, declared to the media in the following days, they too had

92. See Emily Conklin, "Foster + Partners' Marseilles airport extension flagged by environmental agency", August 6, 2019. Architect's Newspaper. Accessed 26 May 2020, https://archpaper.com/2019/08/foster-partners-marseilles-airport-flagged-emissions/

93. See Foster + Partners, "COP25: Foster + Partners launches its sustainability manifesto," 6 Dec 2019. fosterandpartners.com. Acessed 26 May 2020, https://www.fosterandpartners.com/news/archive/2019/12/cop25-foster-partners-launches-its-sustainability-manifesto/

concluded that even if all newly designed or retrofitted buildings would follow green building certification set practices, "global temperatures would still rise by three degrees Celsius." And so, as to limit global temperature increase to the 1.5 degrees Celsius required by the UNFCCC's Paris Agreement, the building industry would have to go beyond "passive strategies," such as solar orientation or more efficient energy management. Architects would have to focus on "using less materials and choosing materials that have less embodied carbon." "With careful analysis and dialogue," Trott concluded, architects might even have to face the absolute challenge of "designing climate restorative buildings."

While assuming leadership and setting a role model, they too at Foster + Partners were perhaps reaching the conclusion that the notion of "sustainability" was, itself, a central part of a sustained collective illusion, which kept us believing that the solution to the climate emergency is to be found and implemented by someone else, in some hopeful, but undefined time in the future.

Even if pledges for carbon neutrality were only a decade or two away, delegates at the COP25 could still comfortably postpone to some next conference a much-needed consensus on essential advancements of the Paris Agreement. Even if signs of climate catastrophe were increasingly undeniable, politicians could still defer any real chance of a Green New Deal for some next election. And even if the *Great Transition* is having a slow start, to say the least, architects and other professions of the built environment can still reliably cling to slightly enhanced, technologically supported and growth-driven notions of "sustainability."

Why? Because in some rosier outcome, there will surely be some unknown force that will manage a miraculous fix for all our troubles.

And this is precisely what specialists involved in climate science and policy have been insistently spelling out as *magical thinking*.

If the climate crisis increases anxiety about the future, it is only human to try and postpone the problems it represents — or to address those problems with a bit of magical thinking. This may happen by way of simple denial of scientific evidences, or by believing that an illusory conjunction of certain silver bullets will provide a solution in due time. In Western contemporary societies, for example, the infinite capabilities of technology or the invisible hand of the market have often taken the place of religion or myth in providing protection and guidance regarding the course of events in the physical world. And, as the field of psychology tells us, magical thinking and its belief in unproven solutions does help reduce anxiety. One may feel more in control, certainly less stressed, once faith rests in some likely answer provided by the latest scientific paper, an optimist economic prediction, or the announcement of a revolutionary technological outbreak.

When it comes to the huge difficulty of decarbonising our societies over the next thirty years, however, it does take some magical thinking to believe that effective solutions are shining around the corner. Even with the recent corona virus pandemic causing sudden and massive behavioural change, it takes some magical thinking to believe that people will dramatically transform their attitudes towards an environmental crisis that always seems too distant. It takes some magical thinking to believe that politicians will unexpectedly break their coalitions with existing economical powers, especially those connected with the fossil fuels industry, so as to trigger the *Great Transition*. It takes some magical thinking to believe that the technological solutions looming in the horizon will provide immediate, economically feasible answers to the enormous task at hand.

With evidence of the building industry's substantial contribution to global carbon dioxide emissions, it is only natural that magical thinking will increasingly permeate its discourses. As an example, it is only natural that we will desperately want to believe that the latest advancements on greener concrete will provide the answer to all of the industry's anxieties. Such news may defy some laws of causality, but we will still cling to them as a much-needed source of hope. Faced with an unpredictable outcome, this is precisely the type of idea that offers the comfort that a silver bullet is actually already zooming in.

With concrete and cement production progressively described as weapons of mass ecological destruction, what better news could we expect than an unexpected technological reversal that suddenly makes concrete a safe building material?

Not surprisingly, by the last months of the 2010s there was a real flurry of reports on budding miracle solutions for cement's "carbon problem." And there was good reason for these to be popping up like mushrooms in a wet forest. As one thorough article hinted at, while most people increasingly worried about the global aviation yearly CO2 emissions, not that many knew that the cement industry alone produced about three times as much of the stuff.[94]

Also, with predictions of increased demand of concrete over the next few decades, as well as expectations of the building stock having doubled by 2060, the problem was only going to get bigger. As such, from start-ups to university research units, and from bigger companies to environmental philanthropists, many were fretting about to figure out how to reduce cement's impact on global warming and the environment.

Researchers and makers were looking for new formulas at every stage of concrete production, from the heat generation needed at cement kilns to the potential alternatives for its curing processes. In the report mentioned above, the executive director of the Concrete Sustainability Hub at MIT, Jeremy Gregory, referred to dubious "cleaner-burning" fuels such as agricultural waste or car tires as an answer to cement production's intensive energy demand. More promisingly, there were news elsewhere that his colleagues were rather investigating how to capture CO2 released in the chemical reactions involved in cement-making, so as to turn this into an emissions-free process. From the same source, we could also learn that attempts at greener versions of concrete had already been proposed over the last years, involving surrogate substances such as "desert sand, nano platelets and human urine."[95] Another report pointed to achievements by startups CarbonCure, Solidia and Carbon Upcycling Technologies, which in slightly different versions all managed to get CO2 into concrete's curing process,

94. See Maddie Stone, "Cement has a carbon problem. Here are some concrete solutions." 20 Nov 20, 2019. grist.org, Accessed 3 June 2020. https://grist.org/politics/cement-has-a-carbon-problem-here-are-some-concrete-solutions/

95. See Rima Sabina Aouf, "MIT researchers develop emissions-free cement production process, 7 October 2019. Dezeen. Accessed 3 June 2020, https://www.dezeen.com/2019/10/07/mit-researchers-emissions-free-cement/

both speeding up solidification, reducing the needed amounts of cement, and, of course, guaranteeing the carbon sequestration into the artificial rock.[96]

96. See Vince Beiser, "Concrete is Awful for the Planet. Clever Chemistry can Help." Wired magazine, The Climate Issue, April 2020.

The problem with all these apparent miracle solutions was, albeit, that they were still far from being a concrete response to concrete's environmental foes. While the doomsday clock is ticking, these proposals are still at the infancy of their scientific research, or very far from being a viable, attractive market alternative.

As some of the current geo-engineering proposals for the climate problem, they are either extremely expensive or almost impossible to scale up soon. Promising paths to deploy renewable energies in the highly demanding cement production may well be backed by millionaire philanthropists, such as Bill Gates. Yet, breakthrough solar devices that can concentrate and deliver the required 1000º C of heat capacity to bake cement's formula are still at an early experimental phase. They are still a good distance away from being deployed at industrial scale. And, unfortunately, they would still not avoid tonnes of carbon dioxide emissions involved in cement's chemical processing.[97]

97. See Daniel Oberhaus, "A Solar 'Breakthrough' Won't Solve Cement's Carbon Problem" 22 Nov 2019. Wired. Accessed 3 June2020, https://www.wired.com/story/a-solar-breakthrough-wont-solve-cements-carbon-problem/

Some of these are steps are going in the right direction. For sure, the architectural world is anxiously looking at such advancements as an escape route to having to completely abandon the heroic narrative of concrete construction. But such miracle fixes still function largely as comforting wishful thinking.

While some researchers and entrepreneurs deploy their scientific knowledge and optimism in further developing those solutions, others, like architects and builders, can find solace in the promise they provide. But a soberer approach would have to recognise that these results are still years, if not decades away from delivering cement's own silver bullet. While these technologies could one day help mitigate cement's impacts, much would have still to be done just simply trying to reduce the centrality of this material in the building industry. When the magic wears off, that need still leaves the architecture world with a concrete problem.

With no reduction in the amount of construction effectively being faced, alternatives to concrete have stayed on the table — even if they, too, imply some forceful magical thinking.

Hempcrete, for example, was briefly hailed as another silver bullet hissing at a better future. When Los Angeles practice MILLIØNS conducted a workshop to build a small hempcrete structure in May 2019, this concrete replacement was being deemed as the "miracle material." As one would expect, hemp absorbs airborne dioxide carbon while it is growing. More surprisingly, it still captures greenhouse gases even after being harvested and deployed as a building material. When mixed with binding agents lime or clay, its fibrous consistency "demonstrated better ventilation, fire resistance, and temperature regulation properties than its predecessor."[98] Being a lighter material, its transport would also require energy, thus reducing it carbon footprint.

In another project, started by UK-based Practice Architecture, a hemp mix was tested for its ability to originate an easy to assemble, pre-fabricated panel system.[99] Yet, its use in blocks or panels precisely hinted at the material's limitations. As news revealed that it would hardly substitute for the structural qualities of concrete or steel, the promises of hempcrete started to fizzle out.

As mentioned in an earlier chapter, timber also now carries the bond of being "the new concrete" — but that does not come without its tolls. As a BBC news piece put it in mid-2019, there is certainly a kind of magic in drawing carbon "from the air" so as to just "store it in our homes and offices."[100] And many factors seem highly favourable: harvesting mature trees may help with forest management, its lightness again implies lesser construction carbon footprints, and, of course, it displaces other carbon intensive materials.

Nonetheless, if a massive move towards timber could forecast a new carbon-free mystic for architecture, some express their doubts on the the industry's ability to avoid end-of-life carbon release when buildings are dismantled, or to effectively use timber as a replacement for concrete at the scale and ease

98. See Shane Reiner-Roth, "MILLIØNS explores the future of hempcrete in the United States," October 15, 2019. Architect's Newspaper. Accessed 3 June 2020, https://archpaper.com/2019/10/future-of-hempcrete-united-states

99. See Natasha Levy, "Hemp is used on the inside and out of Cambridgeshire's Flat," 9 Jan 2020, Dezeen. Accessed 3 Jun 2020, https://www.dezeen.com/2020/01/09/flat-house-hempcrete-practice-architecture-margent-farm/

100. See Tim Smedley, "Could wooden buildings be a solution to climate change?," 25 July 2019. BBC. Accessed 3 June 2020, https://www.bbc.com/future/article/20190717-climate-change-wooden-architecture-concrete-global-warming

Millions + Hempcrete by ISOHEMP

with which the latter is being used today. Further, while timber has almost none of concrete's carbon emissions, it is still based on extraction of natural resources, which at a certain point may itself be proven unsustainable. Especially if construction demands continue to grow at an extreme rate. And, finally, although timber even contributes to carbon sequestration, there is a limit to the amount of land that could be diverted for its creation, especially as such land use increasingly competes with human food production for a growing world population.

Building materials such as these have been repeatedly targeted as potential silver bullets to counteract the high environmental impacts of the construction industry. And in the search for such alternatives, architects have been seen as essential go-betweens to achieve the magic of a positive transformation.

Ultimately, architects are the professionals in a position to be the early adopters of such innovations, and also the pedagogical promoters of such alternatives through their design indications. And while they can certainly help by changing their aesthetic and disciplinary imperatives, it will be the wider workings of a market and industry that will determine if those solutions make it beyond the realm of magical thinking.

Dalston Works, London by Waugh Thistleton.
Photo by Daniel Shearing

Unfortunately, some of these alternatives are also presented as the panaceas that will allow architects to continue their 'business as usual', i.e. go on designing precisely in the same way they have done until now, only trading a contaminating material for its new magical replacement. However, when it comes to a major environmental crisis, we might have to dig deeper than this. Solutions may have to attend to an increasing complexity. Clean and easy silver bullets may not do the job.

There are wider issues in which only policy at the larger urban scale may impose lasting transformations. And, then, there are those problems in which only continued scientific research can determine what ideas may turn from appealing magical thinking to actual solutions to pressing environmental problems.

Take the urban heat island (UHI) effect, particularly as it dramatically intensified because of global warming. In 2018, a New York Times front cover piece announced that summer heat in India's cities could "soon be literally unbearable." One of the ways to address the problem seemed magical, if only because of its simplicity: some cities were painting roofs white. This actually sounded like a handy application of the scientific principle of increasing albedo, or the earth's surface reflectivity, so as to cool the atmosphere. Tests in Hyderabad indicated that this had "brought down indoor temperatures by at least two degrees, which was enough to make the intolerable tolerable."[101]

101. See Somini Sengupta, "In India, Summer Heat May Soon Be Literally Unbearable" 17 July 2018. New York Times. Accessed 6 June 2020, https://www.nytimes.com/2018/07/17/climate/india-heat-wave-summer.html

102. See Jiachuan Yang, Zhi-Hua Wang, Kamil E. Kaloush, "Environmental impacts of reflective materials: Is high albedo a 'silver bullet' for mitigating urban heat island?," in Renewable and Sustainable Energy Reviews, 47, 2015

This could entail a classical return to the sound contextual response of traditional architecture to its harsh environments. However, it turns out that perhaps the magic was not that straightforward. Parallel scientific research was inconclusive about the overall effects of such measures. While one paper recognised that temperatures did seem to lower locally due to the application of reflective materials, it also identified unintended consequences in the surrounding areas, including "adverse regional impacts on the hydroclimate." As the scientists involved in this study concluded, "neither high albedo materials nor any other individual technology, seem to be a silver bullet for mitigating UHI."[102]

As a cautionary tale, this tell us that some concerns must be kept in mind when we warm up to seemingly flawless, inspiring fixes. Scientific knowledge, if always evolving, is indeed uncertain, and constantly subject to revisions. What may look like a fit-all solution at a given moment, can be demonstrated to have unintended consequences some time after.

In the case of some recent geo-engineering proposals, it is only a wild guess what kind of impacts could arise from further tinkering with the planet's climate system. Spraying reflective, polluting aerosols in the upper levels of the atmosphere could effectively help lower the Earth's atmosphere temperature — and thus substitute for more difficult economical and political measures required to counter the climate crisis. There are scientists who defend such solar radiation management as a technological fix to a warming planet. But even those will confess to the utter ignorance of what the unintended repercussions could be. Others see these almost magical, less expensive solutions as an unacceptable hidden excuse to continue emitting greenhouse gases at will. On the other hand, potentially less harmful technologies, such as spectacular looking machinery capturing carbon dioxide directly from the air, come at a painful price — and are still far from being scaleable to growing needs of removing carbon from our planet's atmosphere.[103]

103. See Christa Marshall, "In Switzerland, a giant new machine is sucking carbon directly from the air," 1 Jun 2017. Science Magazine, Accessed 9 June 2020, https://www.sciencemag.org/news/2017/06/switzerland-giant-new-machine-sucking-carbon-directly-air

For now, such magical measures may also seem to be remote from the architecture realm. But are we really that far from having to design tall buildings that integrate massive carbon sucking devices as part of their design facade, or indeed of their primary conceptual strategy? Surely, those would be a novel added feature to the now stale language of high-tech architecture. As proven by solar and wind energy technologies, unlikely scenarios for mass use have been made feasible sooner than expected due to plummeting costs. In time, the spell may actually work.

While we wait, however, we could point out that there are more convenient ways in which other carbon sequestration methods can be brought home to current architectural thinking. Indeed, as already hinted at in a previous section, this is where the allegiance of building design with landscape architecture may prove to be newly attractive as a productive, ecologically driven interdisciplinary interaction.

Technologies to remove carbon dioxide from the atmosphere do not have to be necessarily industrial. They can also be biological. And this is where practices of landscape architecture can typically be influential. Given the right conditions and scale, urban afforestation, will sequester more carbon than any machine will ever do, and at a fraction of the cost. As soon as city authorities make it a mandatory policy, landscape architects have the skills, and hopefully the public ambition, to make it happen. Most biological carbon sequestration techniques have to do with land use and the transformation of landscape, and landscape architects have the knowledge and the creative ability to deal with those in very different contexts.

The possibility to bring landscape architecture's carbon sequestration techniques to the realm of architecture and city-making must surely be embraced. But even those magical prospects should be taken with the necessary precautions. As landscape architect Bill Fleming has recently warned, we must keep a healthy suspicion about the "myth of landscape architects as climate saviours."[104] Only if because of the profession's small numbers and limited influence, there would be a gap between the limited reality of its practice and the ongoing rhetoric about its capabilities to impact the built environment at a large scale — as engineers and architects had done respectively throughout the 19th and 20th centuries. Even if there were historical calls for landscape architects to become "the steward[s] for the biosphere," Fleming claimed, the niche solutions the profession is currently offering end up being mostly propaganda.

104. See Bill Fleming, "Design and the Green New Deal," April 2019. Places Journal. Accessed 9 June 2020, https://placesjournal.org/article/design-and-the-green-new-deal/

Indeed, there may be practical limits to how much landscape architecture's set of tools can be massively adopted as an effective response to the magnitude of the *Great Transition.* However, this does not invalidate the tools *per se*. It only confirms that new disciplinary allegiances could put forward a potential path to expand the impact of the adequate ideals behind them — and perhaps make way for magical thinking to give way to actual results.

Where certain forms of knowledge face obstacles in their direct implementation, propagation to other fields can certainly provide an alternative, if less straightforward advancement of relevant ideas. As Fleming points out, landscape architects "may yearn to impart systems-level change," but they are still "working on discrete sites, with incrementalist tools, within structures that produce injustice." As such, if landscape architecture was to be more transformative, it should be more political, and seek the "tangible support from allies in fields like sociology, law, and science." Similarly, its wider imprint on the built landscape would also require a broader allegiance with architecture practice.

And yet, perhaps this bridging between the two design disciplines has to actually spring from the field of architecture. While keeping in mind that single silver bullets are unlikely, the practice of architecture itself must first realise the potential and accumulated knowledge of landscape architecture's specific strategies for carbon sequestration. While seeking for the direct collaborations that will allow that knowledge to transfer to the realm of building and urban implementation at wider scales, architecture must certainly open itself to a reconceptualisation that privileges organic over mineral. While exerting caution in regards to any remainders of magical thinking, we surely want to welcome any answers that are likely to produce positive outcomes.

In the face of a still widely unacknowledged, but pending ecological crisis, we do have to hope that some magic and a good number of silver bullets accrue — so that we actually start to tackle the challenge ahead. And that implies also learning from the mistakes and the illusions that we will inevitably lodge along the way — as Fleming prompt us to do, when he embarks on a harsh analysis of a highly resonant climate resilience ideas competition that turned into a large failure.

In the United States, on the aftermath of Hurricane Sandy, many proposals for the 2013 *Rebuild By Design* federal competition entailed collaborations by high-profile architects and landscape architects. But, while ideas such as BIG's flood barrier system for Lower Manhattan or SCAPE's ecosystem restoration for Staten Island got a lot of publicity, they also ended in pale versions of their initial ambitions or utterly questioned in their core feasibility. For Fleming, the outcomes did not match "the scale of the climate emergency," nor did they provide for faster and better answers. In fact, while that was the shining promise — and even considering that infrastructure implementation is always slow — none of the ideas got into construction until today.

BIG's flood barrier system for Lower Manhattan

In the projects selected to be developed by the *Rebuild by Design* program, as the big designer's magic touch waned, so too initial intentions of community engagement in the planning were quickly watered down. This lead other critics to point out how these costly initiatives, devised to promote innovation within a neoliberal setting, failed to engage "in the bigger critique of what capitalism is doing and the ways that development is continuing to endanger vulnerable people in cities."[105]And this critique certainly put the finger on a common thread of magical thinking behind many proposed, but ever distant climate actions.
As environmentalist Ashley Dawson said it, when we get desperate, we will put our hopes on anything.

105. See Zach Webb, "An Uneven Disaster: Q & A with Ashley Dawson," The Baffler, 15 September 2017. Accessed 9 June 2020, https://thebaffler.com/latest/an-uneven-disaster.

ECONCRETE UNIT
PORE SPACE FOR FISH
HIGH TIDE
LOW TIDE
EXISTING SEA FLOOR
Breakwater placement avoids hard clam habitat and other critical species.
SUBTIDAL REEF STREET
Subtidal rock enhancements provides structure for juvenile fin fish and lobsters.

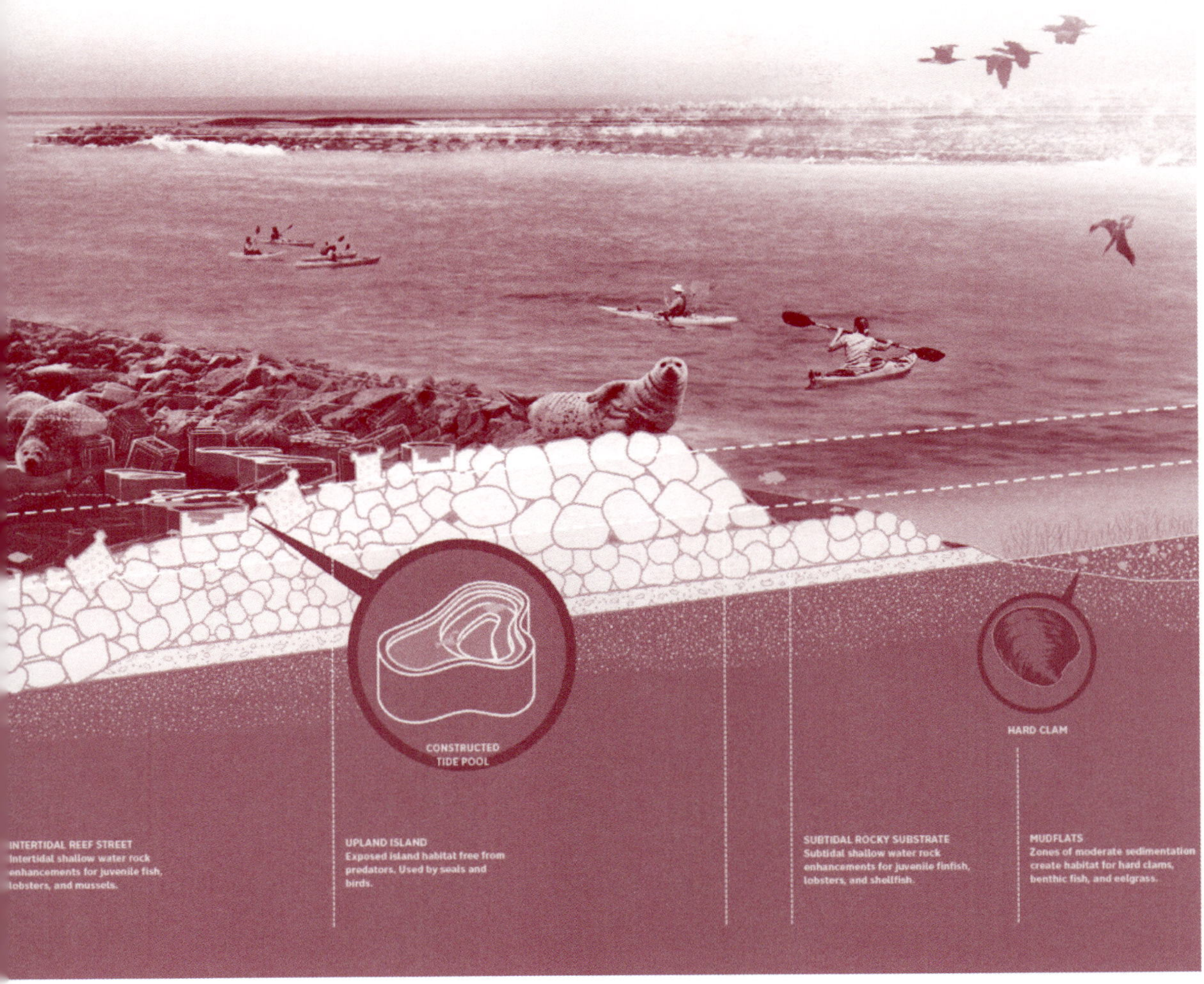

Ecosystem restoration for Staten Island by SCAPE

The risk is that major political endeavours, such as all the Green New Deals we have already touch upon, become the biggest source of magical thinking. That silver lining of combining an impressive environmental transition with economical growth and social justice, while endearing, may simply be too good to be true. That is, in fact, one of the insights of Bill Fleming's analysis of design postures vis-à-vis the Green New Deal. While this may be the "biggest design idea in a century," or maybe because it is so, "no organization has stepped up to articulate the extraordinary scale, scope, and pace of landscape change that is implied."

When looking at the political ambitions of the United States or Europe's Green New Deals, climate scientists are typically suspicious of even the technical capacity to achieve those plans' proposed decarbonisation goals. More so within the limited time spans called out by politicians in electoral campaign mode. Simply put, as many as they might come, for those scientists any prospective silver bullets have not yet proven their ability to meet the demands of the energy transition ahead.

As one author of many pointed out, then there is also the small question of cost. Writing at the *New Statesman* in mid-2019, Paul Mason poignantly asked the questions that really mess with the magic: "who will pay, whose interests will suffer, who will mandate the change, and are the ambitions consistent with democracy?"[106] For architects, as for anybody else, these are the dilemmas that pester the galvanising wizardry of all those Green New Deals. They annoyingly bring us back to the essential doubt about the persistent imbalances of our current economical system. They bring us down to the hypothesis that, while desperately embarking in magical thinking so as to try and maintain the status-quo, we may be missing that most painful evidence: permanent economical growth may, after all, not be sustainable.

106. See Paul Mason "Manifesto for a post-carbon future," New Statesman; London Vol. 148, Iss. 5480, (Jul 19-Jul 25, 2019): 32-35.

As Mason starts, "it was once widely believed that time was on the side of progress." However, because of climate change, we are now running out of time. There is no longer time to wait for some future technology, or the ultimate magical increase in efficiency, to hopefully cover for the costs of a permanent march forward. Besides, those costs now include previously unacknowledged environmental tolls, as much as an increasingly unacceptable social price.

Due to the successive reapings of colonisation, industrialisation and continued resource extraction, enough wealth was produced to raise millions from poverty along the 20th century. We could design and build a brave new world. But now we are seemingly hitting a wall. And we are not hitting a wall because there are no longer resources to be mined or drilled. We are hitting a wall because, as per current scientific knowledge, we now know that maintaining a *status quo* of blind growth equals ecocide and self-destruction.

As Mason and others warn, if no global carbon taxes are quickly put in place, governments will have to incur in massive debt to start any kind of Green New Deal. Yet, as opposed to the magic of past Keynesian episodes of monumental public investment, the real risk at this moment is that "growth, as measured by GDP, will never be strong enough to erode the debt."

As seen during the Covid 19 pandemic, a true emergency implies a centralised government enforcement, disruption of existing key business models, and mass behavioural change, which Mason had pointedly identified with the implementation of any sort of a Green New Deal. Bruno Latour, as others, has asked if the recent pandemic was a dress rehearsal for a larger ecological crisis.[107] In the case of such rehearsal, the global spread of that virus has only demonstrated that, in the face of a looming catastrophe, there is no market system or existing model of "growth" that can deliver the timely solutions for the economic crisis that ensues.

107. See Bruno Latour, "Is This a Dress Rehearsal?," 26 March 2020. Critical Inquiry. Accessed 10 June 2020, https://critinq.wordpress.com/2020/03/26/is-this-a-dress-rehearsal/

Even if sprinkled with some magic dust, a Green New Deal may still be the best alternative for ours and the planetary ecosystem's survival. Then, as many have been pointing out, perhaps it is about time that we just reconsider how "growth" itself is defined.

For Mason, for example, we can start by recognising that the capitalist system was attractive for 250 years only because it had fossil fuels and labour exploitation as "its its twin sources of dynamism." If, while tackling an ecological emergency, we aspire to change both, it is that overall economic model that is brought into question. As he points out, even the concept of a "just transition" may have to be interrogated, as indeed the consequences of a major ecological and economical disruption are unevenly distributed — as, again, the recent pandemic has harshly confirmed. In the end, it always comes back to how we may redefine economic growth, its direct benefits and its unintended repercussions.

When it comes to the implications of *The Great Transition* on our built environment, techno-utopians still believe that, like in other sectors, we may magically decouple innovation and technical fixes from political change and behavioural change. This alone would provide a way out of the climate emergency. They believe some silver bullets will do the job. In the architecture world, some engage in the even more incredible magical thinking that if we look away and stay focused on the aesthetics of form-giving, all problems might in due time vanish. But, if we are going to be honest, we may start glimpsing that neither of these options are truly manageable.

As that early and prescient New York Times article on the Green New Deal hinted at, when politicians too rely on a silver bullet to reduce the world's economic system dependency on fossil fuels and the emission of greenhouse gases, they are simply postponing the sacrifices that would cost them votes already today. As Thomas Friedman wrote at the time, he no longer believed that one "Manhattan Project" would solve the energy transition problem. The previous New Deal had not been built on a magical bullet, but on a broad program of initiatives across sectors.

As he then added, "if we are to turn the tide on climate change and end our oil addiction, we need more of everything: solar, wind, hydro, ethanol, biodiesel, clean coal and nuclear power — and conservation." Similarly, in the realm of architecture, we will also need a new language for the built environment.
We will need new materials, new forms, and innumerable new

ways of practicing. Rather than relying on magical thinking, we need to find the tools and tactics through which we incrementally adapt to a new economic, social and political system, where economic growth, as we knew it, is put into question. Past the false incantations of notions of sustainability still abounding in the field, architects must surely reinvent the imperatives of their discipline's recent history. In case the Green New Deals get going, those would be essential prerequisites so as to fully embrace the opportunities of the coming transition.

Moreover, if prevailing notions of economic growth are also to be overthrown along the way of the environmental crisis, there are other more radical ways in which architecture has to reinvent itself as a discipline and as a profession. In fact, architecture must reconceive itself as a design practice that is able to move from a world view based on incessant new construction, akin to a model of permanent growth, to a new paradigm aligned with a revamped ecological thinking.

The belief in eternal growth in a planet of limited resources may effectively be the most ingrained form of magic thinking. And this is not the sort of magical belief that, while relying on some illusive postponement, at least believes that some transformative change is required. On the contrary, it is the kind of superstitious credo that tells you that if we don't change anything perhaps, just perhaps, nothing evil will actually happen.

In the face of unbearable stress, such as that provoked by the current environmental emergency, we have an all too human tendency to accommodate superstition and make-belief. This is the reason why we have to make an effort to confront some unwanted, but logical options.

In the case at hand, we finally have to make a better effort to accommodate some thinking about that most scary of human prospects: *degrowth*.

8.
Digesting Degrowth

In the winter of 2017, I concluded extensive research for a major curatorial project on the responses of art and architecture to the ecological emergency. This produced a major turn in my state of mind. Shortly after, I found myself in snowing Oslo to be part of an international jury that would choose the curators for a major architecture triennial.

On arrival to Norway's capital, I remember thinking how ironic it was that the Scandinavian countries, which had achieved the most equitable social democracies in Europe, were also soon to presumably boast the most welcoming climate. While the continent's Southern tip was already suffering from longer lasting droughts and extensive soil erosion, in two or three decades the cold countries up North could actually be enjoying a Mediterranean spell of permanent good weather. How was that for climate justice?

Needless to say, the Nordic societies were also international leaders in the preparation for the *Great Transition*. In fact, as Norway exported its abundant oil reserves to other countries, it used the resulting profits to eliminate its dependency on fossil fuels — and make a definitive turn towards renewable energies. Politicians and people seemed aligned in building a greener society. And, as described earlier, leading architecture practices such as Snohetta were international figureheads in the oaths to decarbonise the building industry.

Taking all of this in, the morning the jury proceedings started I felt compelled to make what I thought was a daring suggestion. The organisers of the Oslo Triennale had proposed the future relevance of architecture as the overarching theme for the event's third edition. During the previous summer, I had confronted the magnitude of the ecological crisis ahead. I could no longer disconnect any foreseeable future from its wide impacts. After that transformative epiphany, future architecture could only be relevant if it somehow dealt with the reality of climate change and its reverberations in a wider ecological system. Consequently, I suggested to my jury colleagues, this could be a first filter through which we could look at the seventy-one curatorial proposals we had in front of us.

To my surprise, my colleagues readily agreed to the working strategy. Perhaps because they all came from Scandinavian countries, they had already warmed up to the concerns I had just recently espoused. While I was embracing activism out of shock, when it came to the climate emergency their progressive social context had already made them well aware of the demand for behavioural change on every front.

For the sake of the jury's equity, we did agree to keep on our shortlists any proposal that a single member of the jury deemed as a qualitative response to any other major issue, which could affect the future course of our societies. We did analyse in greater detail proposals that looked at the coming impacts of artificial intelligence or dramatic changes to human labor. But, unconsciously, the path was set for our discussion to primarily focus on the dozen competition entries that, in one way or another, established a compelling nexus to the inescapable climate crisis.

Interestingly enough, when we came to the final round of debates on the last three shortlisted proposals, an unexpected tension entered our conversation. One of the proposals that the jury was attracted to — and the one that eventually would win the competition — was about a topic that left many of us uneasy, including the organisers of the triennial: in the face of some of the inevitable scenarios brought by an environmental emergency, what would an architecture of "degrowth" look like?

For the first time, I witnessed a discomfort that I would see again and again, whenever I had the wild idea to utter the word "degrowth" in the middle of some group discussion. Here, a distinct concern emerged that a major public initiative praising the advantages of economical degrowth might be too disruptive for consensus-loving Scandinavian societies. These were countries which, let us recall, had pulled themselves out of historical poverty only after World War II. To think that, so shortly after, economic growth would be reversed rose heated emotional suspicions. More over, there were worries that the young architects and curators who proposed the theme could be the spearheads of radical international movements calling for dangerous political and economical upheaval. Phone calls had to be put in place so as to make sure this was not the case.

Finally, all doubts were cleared and the team composed by architect Maria Smith and critic Phineas Harper, from London, Canadian architect and educator Matthew Dalziel, and Norwegian researcher and artist Cecilie Sachs Olsen, got the go-ahead to start preparing the 2019 edition of the Oslo Architecture Triennial.

As the curators cautioned in an interview on occasion of the Triennale opening in September 2019, the concept of "degrowth" had not been called upon as a doomsday signifier for social regression or economic collapse. It was rather used to symbolise new avenues for an equitative redistribution of finite resources.

In contrast to the aggressive hegemony of commerce, for practicing architect Maria Smith this was a condition for the viability of "more socially and environmentally conscientious projects." As she added, "if we could free ourselves from this incessant, necessary growth then a lot of those really culturally rich projects could flourish."[108] More than a proposition of loss, "degrowth" would mean changing the terms by which we measure our productivity. As countries such as New Zealand started to consider, the idea emerged that there could be a happy life beyond the rigid horizons of Gross Domestic Product. Confronting how the obsession with GDP carried both social injustice and a heavy environmental toll, human development could thrive on other sorts of climaxes.

As the opening statement for *Enough: The Architecture of Degrowth* had it, "for the last two centuries, the engine of architectural production and the basis of societies around the world has been the pursuit of economic growth."[109] Yet, the climate breakdown was clarifying the limits to that model.

In the team's perspective, architecture had an essential role in such prospects. If redistribution should be happening "by design," and if imagination was required to reinvent notions of growth, spatial practices could have an impact that, even if at small,

108. See Mark Minkjan, "Degrowth is about redistribution by design, not by collapse" 17 September 2019. Failed Architecture. Assessed 23 June 2020, https://failedarchitecture.com/degrowth-is-about-redistribution-by-design-not-by-collapse/

109. See *Enough: The Architecture of Degrowth*, September 2019. Oslo Architecture Triennale. Assessed 23 June 2020, http://oslotriennale.no/en/aboutoat2019

local scales, could parallel that of adequate public policies. In the face of finite resources, both sustainable and "green" growth were revealing their inner contradictions. While they remained politically attractive in ideas of a Green New Deal, they also hid the central problem of the physical impossibility to expand resources forever.

In contrast, ideas of degrowth posited that both the building blocks of the economy and the existing social pacts had to be reconceived. In this context, as much as it was already offering a myriad of potential solutions to rethink everyday building systems, architecture could also provide new spatial configurations for emerging collective arrangements. Again, the climaxes that made the architecture world kick could move into new directions and aspirations.

In the Triennale, this general proposition first translated into exhibition sections that subverted institutional typologies, reenacting a theatre as a revelatory building site, upcycling a garage into an exhibition space from which the city was appropriated through play, or recycling a museum as a library. In the latter, the encyclopaedic approach allowed for a most expansive show of architectural responses to "degrowth." It was here that, upon an international call, many architectural practices revealed how they were exploring conceptions of architecture that could be seen aligning with a different vision of society.

OAT 2019 Visual identity, by Studio Christopher Victor

Norell/Rodhe, Under Construction, 2019

Proposals of design projects already being researched and implemented ranged from entrepreneurial social programs deploying waste to make new products, such as the Oslo Living Lab, to buildings and neighbourhoods made of discarded building materials, by Gusto or Norell/Rhode & Josefin-Wangel; from construction with bio-bricks, in which clay extraction naturally limit the amount of building in a given geographical area, by Jesse McCormick and Khoi Nguyen, to *Another Column*, a textile structure surrogate to be filled with whatever locally available aggregates, proposed by YYYY-MM-DD.

Another Column by Kate Yeh Chiu and François Sabourin (yyyy-mm-dd)

Many architects proposed reversals of design processes, privileging the use of alternative resources, rather than favouring the ongoing logic of extractivism and ecocide. The more than 80 design contributions offered a rich panoply of ideas complemented by critical theatric performances, a documentary on a large rewilding project in northern Norway, and even a revisit of a 1969 exhibition, *And After Us*, which marked an early questioning of a growth-based society by environmentally minded students of the Oslo School of Architecture.

In this diversity of small-scale but inventive solutions we could already glimpse what I here call the one thousand pathways for the *Great Transition*. Rather than the one-size-fits-all solution, or the one overarching architecture language that fuelled the Modern

GISTO, "Multiplo", The Libray, The National Museum - Architecture. Photo: OAT / Istvan Virag

Movement, when confronting the environmental crisis, we are in need of multiple, additive ideas, which can replicate in numerous locations according to specific resources and skills. We not only need our design culture to shift to "place environmental and social concerns at the top of the list of priorities," as Smith put it in the same interview. We also need for it to overcome its obsession with a media-driven international style, rather focusing on community- and resource-oriented solutions and forms.

In this sense, the curators of the Oslo Triennale got it right when, recalling feminist scholar Donna Haraway, they insisted that the focus for these times should be on "maintenance" and "care." As opposed to an economy that has compulsive consumption at its core, we now have the technology and the knowledge to upgrade organisational models that traditionally held a more balanced relationship with environmental resources. From regenerative agriculture to bartering, and from the commons to a full circular economy, by way of accrued scientific knowledge and new digital tools, we now hold the potential to no longer look at these processes as a form of regression. Similarly, we can imagine an architecture that relies on maintenance, rather than on the ever-new display of technologies and materials for robustness and permanence. As Maria Smith put it in quite poetical terms, ongoing care of buildings could become "a good social ritual, a festive act that could make much, much kinder buildings."

From this perspective, the activity of design could indeed transform into a very different endeavour, with Smith calling for a notion of architects as "spatial therapists." By yielding to degrowth, architects would necessarily enter a different relationship to resources made available for their designs. Rather than looking at resources or materials as mere assets to achieve an architectural form that remains indifferent to environmental milieu, architects would rather engage in "negotiating and configuring the existing resources in order to therapeutically benefit the entire context that we're working with."

As another member of the curatorial team added in the same interview, the Oslo Triennale was filled with such propositions, ranging from reexaminations of "older ways of building from earth or thatch" to "ingenious new things like making structures from aggregates but without cement." Many of the projects dealt with recycling or with deploying the greener alternative materials that we have already mentioned here. But, as Phineas Harper warned, these possibilities had to go beyond the mere actualisation of "business as usual" through friendlier means.

"A better version version of growth-based capitalism," Harper posited, "would buy us a bit more time on the planet," — but this would get us only so far without other components. Practices would have to question their contribution to ongoing inequalities and lack of environmental justice in wider city-making or building decisions. And while an architecture of degrowth could only happen in parallel with larger systemic change, the Triennale brought together projects that were already active in inducing such change — be it in housing proposals that entailed new forms of sharing or intergenerational exchanges, or storytelling exercises on scenarios for possible degrowth futures.

One of the most compelling remainders from the Triennale was actually a collection of science fiction stories, which made a funky alternative to the customary exhibition catalogue. Reading those short stories, for example, one could start to imagine how the streets of London would bloom in nature and function if only cars would cease to be our main means of urban locomotion.

As one critic put it, the Oslo Architecture Triennale really was a first attempt from the contemporary architectural field to "come to terms with a growth-free future."[110] In her critical essay about the initiative, Federica Buzzi traced degrowth's history back to the 1970's, underlining how the concept remained under-appreciated and even became a source of distress for people who closely identified it with becoming poorer. As such, while society at large still seems unwilling to recognise that "infinite growth is untenable," it also lacks a predisposition to accept a more sophisticated understanding of degrowth, which also addresses global environmental justice. People, including architects, thus generally miss the important point that, as Buzzi nails it, "degrowth is a collective project aimed at creating an alternative socio-ecological and political-economic order though the purposeful downscaling of affluent economies."

110. See Federica Buzzi, "Coming to Terms with a Growth-Free Future," 26 February 2020. Failed Architecture. Assessed 23 June 2020, https://failedarchitecture.com/coming-to-terms-with-a-growth-free-future/

As many fear, especially in developing countries, degrowth is not about blindingly halting economic growth across the globe. Degrowth could rather be about correcting asymmetries — and eventually adopting leapfrog advancements towards more ecologically friendly development in societies that have not as yet suffered the environmental blow of a full fossil-fuel industrial progress.

And while we are already aware of the magic thinking involved in capitalist economies relying on green alternatives for continued growth, Buzzi noted that the implied transformations in our ways of life would be so massive, that they would still require that we be "able to imagine" what the future would look like. As such, "degrowth in architecture [was] not (only) the use of sustainable materials, and definitely not building harder, better, faster, greener."

For this critic, degrowth would also represent a "different approach to the built environment," with architectures of degrowth "aiming at a radical reform of habits through design, building according to very different collective needs, and even not building at all." It requires us to acknowledge that "the creative industry as we know it (indeed, all industry as we know it) cannot be reconciled with ecological justice." For Buzzi, it would then be indispensable for us to "rethink creativity outside of the growth imperative, not as the privilege of a talented few but as an ingrained human characteristic."

As the Triennale's proposals finally underlined, and as the history of the concept also confirms, degrowth would essentially constitute a conception and a design of a social philosophy for the *Great Transition*.

Not by chance, Austrian-French André Gorz, who first coined the idea of "degrowth," was himself a philosopher. He first spoke of degrowth in the early 1970s, when asking if the earth's ecological balance — which required no growth but only cyclical processes — was indeed compatible with capitalism's permanent thrust forward. Gorz's introduction of the term coincided with the launch of *The Limits to Growth*, the famous report by the Club of Rome that triggered a growing ecological consciousness — and, as described here, effectively influenced many of the environmental avantgardes that emerged in the architectural field at that time.

However, as much as those visionary movements faded, so the disturbing notion of degrowth got deeply buried in the architectural unconscious. Adhering to the market-oriented ideologies that became predominant after the 1980s, architects too became devotees of ever more new construction — both for economic subsistence and disciplinary legitimation. Degrowth, if at all acknowledged, became a synonym of that other subversive idea of stopping all new construction. It became a scary concept, which went unnervingly against the grain of the profession — even if traces of that conception were once a trigger for design innovation.

Indeed, one could assert that, historically, the modernist quest for a minimum dwelling came close to an idea of degrowth. Housing schemes were redesigned and spacial resources reinvented so as to provide access for the many. This was a model that, while responding to population growth, provided a more balanced redistribution of limited assets. If the minimum dwelling responded to social needs at a given moment, it also provided for architectural research to flourish in new ways. However, as already indicated, ideologies changed, and it soon became more attractive to respond to market forces. Architecture practices sweetly subsumed to the expansive logic of capitalism and consumerism. In fact, as it happened in every other social sphere, alternatives were no longer available for thought experiments on redistribution.

After the 1980s and Margaret Tatcher's famous dictum of *There Is No Alternative*, it became lastingly difficult to consider degrowth as a plausible option for *another* social and economic organisation. And, with a broader political development of degrowth, came the inevitable evidence that many people feared and rejected degrowth precisely because they could not foresee it as an alternative to capitalism. Even if increasingly uncomfortable with the negative impacts of the capitalist regime as we endure it today, from blatant environmental disruption to persisting socio-economic inequities, most failed and fail still to see *any* alternative to such a system.

Yet, especially after the Great Recession of 2008, there were rising movements that, in parallel with the dominant public narrative, called for degrowth as an alternative model for society, based on cooperation rather than competition.

A decade ago, one of the most active voices in these movements, economist Giorgos Kallis, precisely acknowledged the resistance that degrowth usually elicits, clarifying some of the assumptions and challenges surrounding proposals of degrowth. Kallis thus defined what he deemed "sustainable degrowth" as "a socially sustainable and equitable reduction (and eventually stabilisation) of society's throughput," referring to the flow of materials and energy that a society "extracts, processes, transports and distributes, to consume and return back to the environment as waste." Importantly, however, degrowth was also presented as an umbrella keyword, from within which a "full ensemble of environmental and redistributive policies" addressed issues such as basic income, reduction of working hours, economic decentralisation and redistribution, environmental taxes, decarbonisation, and many others. With these measures, usually being taken as "harmful" for the economy, degrowth faced outright rejection from the public. As such, degrowth became also "a radical political project that offers a new story and a rallying slogan for a social coalition built around the aspiration to construct a society that lives better with less."

Indeed, with the ecological inevitability of degrowth, the question for this researcher remained only if the process could still be "socially sustainable," or if it had to go through a "catastrophic descent."[111] For the former option to become a reality, we would have to go back to a notion of "selective degrowth," as advanced by Serge Latouche, another important thinker of degrowth. By planning where degrowth should be first put in motion, rather than blindly imposing a single solution that would be indifferent to contexts of development, we could hope to achieve a more refined balance between different economies, different institutional actors, and even different generations.[112]

Consequently, for Kallis, social transformations encapsulated in the call for degrowth do not have to work as direct appeals to those in power, "but by bottom-up movements that challenge established paradigms." This locates degrowth in the dominion of new tactical approaches to the built environment, and connects it to parallel developments in the renewed relevance of the commons, of associative and collective action, and even of

111. See Giorgos Kallis, "In Defence of Degrowth," Ecological Economics 70 (2011) 873–880, Elsevier: 2011.

112. See Serge Latouche, Farewell To Growth, Polity Press, Cambridge: 2009.

bottom-up forms of urbanism. It speaks of degrowth's "fragmented and diverse" nature, but also, because of that diffuse character, of its ability to infiltrate the political and economic discourse in non-conventional ways.

Even if academic discussions around degrowth grew steadily, one could easily grasp that such a movement would gain traction and lose its fringe character only when influential mainstream outlets would dedicate long essays to the topic. With degrowth, that crucial moment happened in February 2020, when the New Yorker magazine decided to finally ask "Can We Have Prosperity Without Growth?"

Investigating a discussion that was "gaining widespread attention in the face of the climate crisis," writer John Cassidy not only described the movement's history and identified its main thinkers, but reported also on why a stagnant economy could actually prove not to be utterly frightening — and be instead a sign of success and full development. In fact, without resorting to reasons of economic inequality and climate crisis, some economists even defended that "slower growth is appropriate for a society as rich and industrially developed as ours." After all, where population is ageing and reducing, one should also expect a decrease in GDP.[113]

113. See John Cassidy, "Can We Have Prosperity Without Growth?" 3 February 2020. New Yorker. Assessed 23 June 2020. https://www.newyorker.com/magazine/2020/02/10/can-we-have-prosperity-without-growth

Japan represented, in this context, the typical example of an advanced economy that, while facing deflation for more than a decade, remained a thriving consumer society and a focus of economic development. Indeed, if architects would need strict reassurance that lack of growth is not a threat to innovative building practices, they would only have to look at Japan's burgeoning and influential architecture dynasties — with one of those even being the subject of the the most visited architecture exhibition in the world in 2016, MoMA's *A Japanese Constellation*. While often focused on small-scale projects, minimum dwellings and the permanent upcycling of the urban fabric, the contemporary architecture scene in Japan proved that the absence of robust growth was no impediment to design creativity, and,

House N by Sou Fujimoto.
Photo by Iwan Baan

on the contrary, it even bolstered the unexpected refinement of everyday architectural solutions for fully packed cities.

In reality, during its decade of economic stagnation, Japan rather offers a window into the potential paradoxes of near degrowth. While the general perception was of a slow recession, the country's GDP or CO2 emissions never stopped growing, albeit at rates well beneath other developed countries. And yet, one would think that, other than the expression of effective degrowth, Japanese society during this period was more readily an expression of *accelerationism* — another contemporary thought experiment, which is but a perfect polar opposite to degrowth. Together with *ecomodernism*, this is one of two current tendencies that affirm technological progress itself as an overall silver bullet, which "will allow us to address climate and ecological breakdown while also dramatically increasing production and consumption."[114]

While critics point to the disconnection between *accelerationism* and any semblance of scientific sustainability, they also must warn about the risks of its techno-fix hype for "climate chaos" — as, contrary to degrowth, these movements are unsurprisingly growing in popularity. Although scientific evidence warns us that increasing consumption, energy production and greenhouse gases means crossing and disrupting planetary boundaries — the biological, chemical and physical conditions that have made Earth suitable for life — it somehow seems to be easier to believe that accelerating growth will provide a magical solution to all of our problems. *Accelerationism* may equate the risk of "causing abrupt and irreversible environmental changes that threaten the stability of Earth's systems and human civilisation." Even so, for those whose lifestyles are thus far relatively unaffected by the climate crisis, we still seem to be more thrilled by the climaxes of growth.

As the New Yorker essay reminds us, it would however be unlikely that economic growth could continue to increase at its recent rates, implying that by the end of the century global GDP would have "risen fiftyfold or one hundredfold."

114. See Joanna Boehnert, Simon Mair, "Techno-Fix Futures will only Accelerate Climate Chaos – Don't Believe the Hype," 31 October 2019. The Conversation. Assessed 7 July 2020, https://theconversation.com/techno-fix-futures-will-only-accelerate-climate-chaos-dont-believe-the-hype-125678

As such, the argument goes, we really need a reconception of what growth means. And we need to design new understandings of how growth is measured and enacted, so that it also echoes understandable environmental limits and a much-needed social justice. With economical growth moving beyond its primary goal of lifting people out of poverty, and now having its benefits exclusively captured by a small elite, the ecological distress of exhausted resources could, in fact, be accompanied by social disaster.

The rising perception of such combined disasters was certainly one of the reasons why, suddenly, a plethora of Green New Deals became politically attractive. While avoiding having to face degrowth, and straining to maintain an alignment with the economic *status quo*, politicians could ascertain degrowth's cousin, "green growth," as being "strong, sustainable, balanced, and inclusive." With this miracle solution in view while entering the *Great Transition*, some economists took it for granted that it would be simply counterproductive to stop growth. Improvements in general efficiency, it was argued, could continue to sustain green growth for a while. However, scientists do know that there are known physical limits to how much efficiency you can attain in any given process, material or energy form. So, the fact that the numbers and promises behind all the Green New Deals remain illusive, inevitably brings us back to notions of degrowth as a potential alternative to capitalism's growth model.

Thus, some scholars have started to characterise degrowth as a thought niche that, indeed, could gain momentum as a viable tool to dispute and provoke alterations to capitalism quintessential drive. Even before the Covid-19 pandemic added a fatal blow to the global economic *status quo*, researchers were already positing how several interconnected crises were destabilising the illusion of perpetual growth feeding the capitalist system. This, in turn, could open the way for certain niche discussions to gain traction — and win that wider following that the New Yorker article came to confirm.

Academics in the field of ecological economics, for example, recently spoke of a "degrowth transition" and advocated the notion of degrowth as one of those few emerging visions of the future that "recognise the unsustainability of our current societal trajectory and offer alternatives to the current system."[115] As Vandeventer et al. dissected in a recent paper, capitalist growth and its "extractive drive" are increasingly associated to environmental degradation, but also to persistent inequality, extreme wealth concentration and a myriad of social challenges. With this aggressive progression, conditions are again ripe for alternative ideas of social organisation to dispute the hegemony of the prevailing must of economic growth.

115. See James Scott Vandeventer, Claudio Cattaneo, Christos Zografos, "A Degrowth Transition: Pathways for the Degrowth Niche to Replace the Capitalist-Growth Regime," Ecological Economics, Volume 156, February 2019, Pages 272-286

116. See Latouche, ibidem.

Within the logic of frameworks of change analysed by these authors, ideas of sustainable development or the green economy also compete to offer possibilities of transformation, but they are closely aligned with the existing growth model.

As they further adapt to mainstream needs, they also lose ability to offer themselves as effective alternatives for a transformation of the growth paradigm. While, for example, the notion of sustainability spread to different arenas, its interpretation became so open that it sometimes risked becoming meaningless — and this, in turn, degraded its transformative potential. As core problems persist, and as the landscape of the environmental crisis exerts more pressure on capitalism's "growth fetish,"[116] the researchers suggested, doors will eventually open for other niches of radical innovation to pull through. And this is where degrowth comes in.

While degrowth seeks to undermine the idea that a pleasurable life requires the hegemonic illusion of perpetual growth, it also "exhibits no urgency to define a single, unequivocal vision of the future." This, as Vandeventer et al. propose, may prove to be a competitive advantage when the system of growth really starts to collapse.

And while "degrowth is not presently well-developed enough to replace the incumbent regime," for these scholars it already shows a dynamic that is both symbiotic with other pathways for change, and in itself sufficiently diverse — in the long run making degrowth a candidate with the potential to replace the existing system. In fact, as practices of degrowth usually rely in a multitude of micro-strategies often already tested locally, these can be emulated, reenacted and scaled up to other locations without having to become a univocal ideology.

This aptitude for practical diversity, which is imbedded in the very notion of degrowth, was precisely what emerged at the Oslo Architecture Triennale. After all, the idea of degrowth did not equate with giving in to an inevitable regression. On the contrary, it was a call for action requiring the imagination of alternative tools. As Triennale curator Maria Smith put it, such a move, accruing many smaller actions and many smaller strategies, could add up to "a profound change in our social and economic systems from industrial production and material consumption to cultural richness and social justice."[117]

117. See "OAT 2019 Curator interview: Degrowth and Its Relevance to Architecture." Oslo Architecture Triennale. Assessed 7 July 2019. http://oslotriennale.no/en/news/the-curators-explain-degrowth-and-its-relevance-to-architecture

This intention, as one critic had it, was certainly ambitious. As Eleanor Beumont wrote at The Architectural Review, you had to be brave "to peel back the skin concealing the ugly ribcage of our economic system, its guts ingesting gas, coal, trees, animals, minerals, water and clean air and flatulently defecating an endless stream of clothes, plastic bags and neat packets of processed food." But, while cracking open capitalism's growth system, the curators of the Triennale were also pointing to architects' misunderstandings when considering that they could "confront the climate crisis by merely rethinking the way they design buildings."[118]

118. See Eleanor Beaumont, "Enough is enough: Oslo Architecture Triennale 2019", 14 October 2019. Architectural Review. Assessed 7 July 2020, https://www.architectural-review.com/today/enough-is-enough-oslo-architecture-triennale-2019/10044805.article

Ultimately, Smith left us with the fundamental question of how far can we imagine and start designing a society without growth that, simultaneously, could "radically improve our lives." Other reviewers of the Triennale, such as Matt Shaw at the Architect's Newspaper, duly added many more interrogations, such as how could degrowth be a helpful ideology for architecture, what power structures in the design world could be subject to degrowth, or how could the development and real estate industry be convinced to participate in it.

But, more interestingly, Shaw was left wondering if "every aspect of our lives can be redesigned through the lens of degrowth." He was driven to ask if degrowth could "provoke new ways of building at the individual level that can become communal and then translate into change at the systemic scale."[119] He was triggered to ponder whether degrowth could be just another driver for a climax change in architectural practice. It could just be an eye-opener to the many single trajectories that the architecture world will have to adopt, if it wants to participate in the design and implementation of the *Great Transition*.

119. See Matt Shaw, "What is the architecture of degrowth?," 4 October 2019. Architect's Newspaper. Assessed 7 July 2020. https://archpaper.com/2019/10/2019-oslo-architecture-degrowth/

In the context of the Triennale, these questions revealed only that we do need one thousand pathways, or one thousand architectural tactics, or one thousand spatial micro-strategies, if we want to feed any hopes that architects will remain relevant and will somehow contribute to tackling the magnitude of the transformations that are steadily coming our way.

9.
One Thousand Pathways

If you usually surf the news of the day you have already captured the scent of the latest and truly urban trend. Be it in the first, the second, or the third world, eco-autarchic towers are being devised for the thousands, and sections of cities themselves are being developed as huge private condos for the hundreds of thousands.

There they are: efficient huge buildings and whole city slices growing one after the other, from the deserts of Dubai to the dense grids of Western and Eastern metropolises — as if suddenly stepping out of forgotten futuristic scenarios onto the dirty realism of a gritty, homogenised worldscape.

Even amidst apparent financial arrest, pieces of a technological utopia are rising around us to enter the imagination of the everyday banal. What had been once restricted to specialised visionaries is now the subject of Guinness World Records and a matter of gospel TV. Emphatic documentaries feed us with the amazing features and the heroic characters behind accomplishments in height, dimension and efficiency. But as skyscrapers bloom and new instant cities flourish, one may sometimes wonder what will be the ultimate destiny of these again immensely fashionable mega-structures.

If you are prone to dystopian tendencies you can't but start imagining these constructions like last resorts for the chosen few, as the dejected many riot and turn suburbia and feral city centres into the announced urban slums of the near future. In a fictional mix of Buñuel and Ballard, we could easily picture these enclosed, autonomous structures soon being cut off from the surrounding debris. It's almost as if, knowingly, technology has indeed been thoroughly orientated to create this specific building autonomy for the world to come.

Architecture's aesthetic autonomy may have been many times criticised as the perpetrator of a social divide. But this is really nothing, if compared to the message enclosed in the possibility of real physical autonomy. What we read today as the optimistic promises of ecological efficiency may as well provide tomorrow the perverse solution for the ultimate social segregation between rich and poor.

The harvesting of rainwater and humidity, the workings of superlative solar and kinetic energy recycling systems, the agricultural exploration of hanging gardens, the filtering of pollution through the building's multiple skins — you name it: they are all wonderful tools leading us towards a brave new autonomy. Moreover, get these together with rooftop heliports and high-tech wi-fi communication systems and you will achieve that autistic sense of perfect social wellbeing that today is already possible in otherwise profoundly unbalanced cities like São Paulo.

If the disintegration of regional and federal nations indeed takes place, maybe it is not really city-states what we are again looking into. We may rather be envisaging new concentrated architectural typologies that adequately fit the coming social organisation. Whether as a surrealist update of the horizontal fortress firstly envisaged by Le Corbusier, or as a cynical vertical remix of Fritz Lang's Metropolis work divisions, we can already extrude the physical forms of the post-contemporary from current conditions.

And until these architectural structures are also besieged and taken over, we may be then looking at the brief return of the building-state itself. As such, it is only strangely appropriate to call these eco-skyscrapers the new castles in the air. [120]

120. See Pedro Gadanho, "Taken to Extremes," in Pedro Gadanho (ed.), Beyond No1: Scenarios and Speculations (Amsterdam: Sun Publishers, 2009).

As I depicted in this piece of writing from shortly after the 2008 recession, there have been many scenarios through which our society has tried to picture its transition to a new environmental and energetic paradigm. This was but the dystopian version of how trailblazing environmental technologies in the building realm could be of service to an increasing social and economic inequality.

This, however, was just a peek of a pathway into the future, which we probably want to avoid. And so, we should be looking at other possible pathways, which may serve our society in more just and equitable ways. And they are already here.

Scientists in various fields, for example, have been busy repeating ever more sophisticated mathematic models of the Earth system and society's future developments, precisely searching to identify and characterise pathways and options for our common decisions. Inputs for these models are constantly tweaked to reflect new knowledge, data and probabilities. Each result and each variation originate a set of conditions, which researchers translate into another potential pathway for change. When speaking of scenarios of global warming, sea level rise and other features of the unfolding climate crisis, several possible pathways of action reflect what could or would have to be done to achieve certain goals — like keeping atmosphere temperature increases under 1.5 or 2.0° C above pre-industrial levels for the next thirty years.

Climate scientists, in particular, posit different pathways through which we could achieve decarbonisation and, eventually, a stabilisation of the Earth's climate system. From 'business as usual' to reduction measures implying alterations to our ways of life that are still politically perceived as unacceptable, scientists try to distinguish the consequences of slightly variable specifics. They collect and translate the data that characterises the new geological epoch of the Anthropocene, so as to anticipate the road ahead.

Others, from politicians to technological entrepreneurs, from journalists to philosophers, or from policy-makers to designers, then struggle to cope with such scenarios and fit them into everyday practices and discourses. Some reflections, like the micro cautionary tale above, are grim and dystopian, warning us of the increasing distortions introduced by the unequal distribution of technological solutions to problems that, they too, affect some more than others. Other visions strive to look for opportunities and positive change within reach, hoping that a multitude of incremental transformations will eventually be adopted at bigger scales and add up to a significant shift towards the often-quoted *Great Transition*.

While keeping in mind Italian philosopher Antonio Gramsci's "pessimism of the intellect," the latter possibility is where this book sees architecture contributing with the optimism of its proven will. As much as modernist architects led an authentic revolution in the language and technologies of architecture for the 20th century, many of our contemporaries have already been investigating the applications of current scientific knowledge to the ecological improvement of architectural structures and the built environment. Multiple pathways are already being explored towards a new architectural paradigm. In order to produce a true climax shift in architecture's system of aspirations, however, it is necessary to make these pathways known and to stimulate their adoption, replication and improvement by many more practitioners.

Rather sooner than later, we should start to map a cartography of change where the myriad of possible pathways towards a new environmental balance can begin to coalesce around a few major categories of action. Even if many approaches and ideas remain to be discovered, and even if we miss some relevant examples, it is important to establish an expanded chart of the arenas where we can expect change to affect and shape architectural practice in new directions.

By way of such cartography, we are not only aiming to situate the practices of degrowth described in the previous chapter, as much as they overlap with many of the following categories. We are rather referring to a wide scope of architectural possibilities ranging from the micro-scale of the eco-efficient hut to the expanses of urban redesign, and from the most basic energy reduction processes to the most complex forms of ecological regeneration. We are talking about improved technological and green efficiencies in materials and construction, but also of the most traditional forms of retrofitting and recycling buildings and cities. And we can reach from the individual efforts toward the management of scarce resources, to more ambitious notions of a circular economy or the re-enactment of commons and shared environmental infrastructures.

Besides the many projects already mentioned in *Climax Change*, this chapter brings together a large number of other architecture ideas, which are already foregrounding the radical change that the field must soon embrace as a whole. Going deeper into examples that illustrate each major category of action, this quick survey demonstrates the notion that, even within some sort of unified theory, but differently from the modernist orthodoxy, many diverse pathways are required to embrace the scope of change that the climate crisis implies for architectural practice.

In their diversity, and in their new tastes and textures for a revamped architectural language, these pathways add up to the possibility of a paradigm shift for architecture in response to the ongoing environmental crisis. Yet, they also speak of moving from the grand narratives of modernity — progress, growth, technology, big development, massive scales of intervention — to the imperative to lessen our footprints in different forms, from the reduction of energy consumption or carbon imbedded in buildings, to the revision of wasteful construction procedures or the decrease in the use of non-renewable natural resources.

In this direction, yet another take on the projects included in the Oslo Architecture Triennale led one writer to compile a series of measures that revealed how architecture could "support a degrowth economy."[121] These included known recipes for how formal concepts could involve found materials — from demolition sites to the use of industrial waste that Rural Studio once pioneered — but also of how designs could reject brand new materials and revolve around recycling, from birth to the afterlife of buildings. Other much discussed solutions involved finding ways to minimize toxic materials such as concrete, or to leave no traces at all when a transitory structure would be dismantled. Bolder transversal proposals for reduction by design bridged from "generating social activity around shared economies," as found in Henry Lefroy Brooks's social design for a Civic Laundrette, to the invention of "sustainable forms of tourism," as proposed by Dekleva Gregorič Architects and students from London's Architectural Association.

121. See Amy Frearson, "10 ways architecture could support a degrowth economy," 19 November 2019. Dezeen. Assessed 7 July 2020. https://www.dezeen.com/2019/11/19/degrowth-architecture-oslo-architecture-triennale-2019/

These proposals revealed an array of directions that variously intersect the broad categories of action that this chapter hopes to delineate. Two specific projects in the Triennale, however, could be seen as straight examples or pathways in the first category that we want to enumerate here: *architecture contributing towards fossil energy reduction.*

Civic Laundrette by Henry Lefroy-Brooks at the OAT. Photo by Annar Bjørgli / Nasjonalmuseet

London-based studio Public Works, in collaboration with students from the Oslo School of Architecture, set out an installation, *Power Plant!*, in which a bench in public space was passively heated by compost. Canadian architecture practice Lateral Office, on the other hand, developed a series of designs for zero-carbon infrastructures that harvest power and resources from their own activity and surrounding conditions, including a compost station, cooperative live-work spaces, a community hub that is also a market hall, or a water exchange supported by aquaculture. The two approaches underlined a search for alternative ways of fuelling both buildings and social activity. Yet, they also spoke of an urgent need to lessen the massive energy consumption that our built environment now implicates.

And it was not always like this, of course. Only fifty years ago, global energy requirements were 70% lower than today. And built architecture certainly contributed to such figures. As one architect looking at that history put it, before air conditioning became a widespread feature, "all architecture was by necessity specific to its climate and place." In his call to recuperate and scale "regionalism," Stephan Kieran was on point reminding us that, once, "there was no choice in temperate and tropical climates but to seek a regionally informed architecture that optimized comfort using passive systems."[122] Ultimately, it was air-conditioning that allowed modernism to become an international style indifferent to context.

122. See Stephan Kieran, "Scaling Regionalism," in Dan Willis, William W. Braham, Katsuhiko Muramoto, Daniel A. Barber (eds.), *Energy Accounts: Architectural Representations of Energy, Climate, and the Future*, Routledge, New York: 2016

As Kieran hints at, redirecting design innovation and form-making to the shaping of an energy-passive architecture — often around and against comfort regulations that enforce artificial refrigeration and heating — may prove to be one of the pathways that puts us on track for significant energy reductions. Further, if we really want to kick off a new paradigm, reinventing the language of energy-passive architecture, while simultaneously adopting technological innovations from renewable energy sources, could provide for one of those design climaxes we are so much in need of. We could call this the passive-aggressive approach to new formal pinnacles.

Further, as we add cheaper renewables to the energy mix, many fear that instead of reducing energy consumption we will only aggravate it. Switzerland-based 2000-Watt Society has calculated that, so as to maintain carbon dioxide emissions at a level that the Earth system can deal with, we would need to reduce individual energy consumption to one-third of what we deploy today. But, instead, global energy use is on the rise. Similarly, in the built world of architecture, despite environmental concerns and certifications such as LEED, energy consumption is also increasing. Kieran suggests that, while "a progressive and environmentally responsive form-making needs to proliferate rapidly," a low-cost and easily replicable construction featuring "effective responses to region and climate" could be transformative.

Nonetheless, architects will have to go farther, quickly integrating technological breakthroughs as they come. This can include yet unforeseen net-zero capabilities. But it may also simply mean adding now mundane technological advancements such as improved solar and wind generation, heat pumps, insulation efficiency and many others.

Fortunately, recent experimental projects have illustrated how such pathways for progress can be tested and built, leading to eventual replication and mass adoption. And these become particularly interesting when they go beyond conceptions of energy reduction and, entering a new pathway category, they also offer glimpses of *architecture becoming self-sufficient and helping reinvent grid-dependency*.

We have previously mentioned how autonomous cabins have been popping up in and out of the architecture discipline. Yet, with *Diogene*, a research project by Renzo Piano Building Workshop in conjunction with Favery&Milan and Transsolar, we move into another level of techno-aesthetic joys. Here, the primeval hut intersects the idea of minimum dwelling to become the self-sufficient technological pod of the future. Reinventing the dream of a minimum footprint for human existence, *Diogene* is no longer only dubiously based on the austere spatial metrics and bare functionalism of yesteryear, but also now relies on minimal resources and energy requirements. As described by the Transsolar team, this is a more elaborate thing: "an intelligent dwelling with technical installations and systems that enable self-sufficiency in different climatic conditions."[123] Naturally, the project had to feature the latest in water collection, solar power with battery unit, waste water treatment and compost toilet, heat pump and a sophisticated mixed ventilation system. Significantly, however, it also delivers the futuristic aesthetic and the architectural legitimacy that may make it appealing as a model to replicate in different contexts.

As we speak, however, this pod would surely be out of reach of the average family's budget even in more affluent societies. One easily guesses it is still a rather expensive laboratory experiment.

123. See Matthias Schuler and Anja Thierfelder, "Five Elements." 6 Sep 2019. E-flux Architecture. Assessed 27 June 2020. https://www.e-flux.com/architecture/overgrowth/221632/five-elements/

Diogene, Weil am Rhein, 2011-2013 by Renzo Piano Building Workshop. Photo by Julien Lanoo

Hanwha HQ, Seoul, Zuid Korea by UNStudio. Photo by Rohspace

Even so, it contains the seeds for building applications that, in the near future may effectively see architecture innovation curtail our reliance on major energy distribution grids.

Self-sufficient architecture does not have to be about retreating into the forest. It can rather signify maintaining quality of life by way of localized renewable energies and resources, while still benefiting from the collective benefits of urbanized areas. This may translate into an architecture form conceived to maximize rooftop solar, or even façade solar, as in a recent proposal from celebrated UN Studio.[124]

124. See Antonio Pacheco, "UN Studio brings a responsive solar skin to an existing office tower in Seoul!" 8 May 2020. Archinect. Assessed 13 July 2020 <https://archinect.com/news/article/150196852/un-studio-brings-a-responsive-solar-skin-to-an-existing-office-tower-in-seoul>

As plenty of contemporary examples confirm, and as already described here, self-sufficiency may simply imply that buildings will be conceived as energy producers themselves.[125] Or it may signify the design of how communities organize around micro-grids or alternative energy sources. But, surely, there is ample scope for architectural creativity with an environmental purpose.

The same applies to how we can see *architecture engaging with forms that promote the better management of scarce resources.* Some of the examples mentioned above, from civic laundrettes and compost stations, to Diogene's multiple collection and management systems, already reveal the amplitude of physical and non-physical resources that we see intersecting any architectural conception. From structures that prompt resource sharing and new collective behaviours, to the decrease in use of natural resources, and from waste management and recycling to actual energy production, there are many avenues through which architecture can reinvent the presence of environment-friendlier "resources" in both its conceptual framework and its concrete design proposals.

Building resources may first of all be locally sourced, thus reducing the carbon footprint of their transport for construction purposes. But their use can also attend to availability, possibility of renewal and the planning of adaptive reuse. Naturally, these decisions start at the drawing table. And this implies that architects thoughtfully consider the formal options and construction techniques they are deploying in each given location, each economical context, and each community profile.

In the case of a 2018 project in South Africa, the Langbos Children's Centre by Jason Erlank Architects, such design logic meant a non-profit collaboration between architects and diverse manufacturers and service providers — from Scribante Concrete and EcoDomes Africa, to Sika and AutoDesk — so as to produce low-cost, context-fit structures, which both echo past styles and present trends.[126]

At first sight, this school echoes traditional forms of local architecture. The progressively inclined could even claim this ran counter to true design innovation. Yet, the project's experimental and didactical character instead feeds on concepts and techniques first explored by the environmental avant-gardes with which we started this manifesto for global change. Even if the Langbos's domes receive a common plaster finishing, their entrails are made of plainly sourced, bagged sand, rather than your typical cheap cement block. Further,

125. See Julia Daudén, "People Power: 19 Public Buildings that Generate Renewable Energy" 17 Feb 2020. ArchDaily. Accessed 13 Jul 2020. <https://www.archdaily.com/933204/people-power-19-public-buildings-that-generate-renewable-energy/>

126. See Hana Abdel, "Langbos Children's Centre / Jason Erlank Architects," 22 Jun 2020. ArchDaily. Accessed 13 Jul 2020. <https://www.archdaily.com/942147/langbos-childrens-centre-jason-erlank-architects/>

the local Langbos community was involved in the construction process, and directly learned how to deploy solar power, readopt water harvesting or use building processes such as Superadobe, the long-lasting heritage from California-based EarthCal.

As an African saying goes, real help is about distributing fishing poles, rather than offering fish. While this project recalls that international cooperation must follow that wise lesson, it also tells us how, when it comes to scarce resources, the reinvention of tradition really can happen with a little help from green technological advancements and interdisciplinary collaboration.

As we have heard from architect Mike Reynolds, this is the moment in history in which ideas of alternative processes based on low-cost resources and even waste can be fully accomplished.

In the 1970s, Reynolds' notion of an Earthship Biotecture based on recycled bottles and tires could still sound a bit wacko for the architectural mainstream. Now, however, this is the time in which waste streams are being industrially transformed into new materials as an everyday practice. Reynold's "biotecture" may now be revisited as an expression of climate justice,[127] but it is also an inspiring model for a radical reconsideration of what we now consider noble materials. Given recycling's technological progresses, architecture and urban design have only to quickly catch up with new sorts of specs — and derive some ethical and aesthetical thrusts from it.

127. See Chelsea Schelly, *Dwelling in Resistance: Living with Alternative Technologies in America*. Rutgers University Press, 2017.

This points to yet another major pathway, one which sees *architecture deploying recycling, retrofitting and adaptive reuse as major forms of disciplinary affirmation*. Instead of getting its kicks from brand new constructions, architecture can still derive its creative pleasures from novel forms, inventive conceptions and brave new functionalities — all of which are, however, now superimposed onto existing buildings and recuperated materials.

Perhaps because new crops of younger internet readers are increasingly aware of the environmental crisis, this pathway has certainly become a trend in the world of online architecture media. Newsletter listicles now enumerate the best ways in which architects can promote adaptive reuse,[128] or how recycled materials can be applied in their practices.[129]

The popular ArchDaily web platform effectively made recycling its topic of the month in July 2020. Considering their immense, youthful readership, this is undoubtedly a good start.

It is key excitement for younger architects to know that, thanks to recent technological developments, they can finally deploy recycled tires in playgrounds, roofs and bricks, waste plastic in urban furniture and pavements, or construction and demolition garbage in concrete manufacturing. But it is also significant that they can learn by example from the recognized master builders.

Along these lines, when Pritzker-laureate Wang Shu and his wife Lu Weniu, founders of influential Beijing-based Amateur Architecture Studio, have their work scrutinized from the perspective of its recycling practices, we have a slight shift of the climaxes that may drive the architectural profession. That they respond to a massive transformation of the Chinese urban landscape by taking advantage of the debris originated in countless demolitions, can become a significant source of inspiration. That their appreciated architectural forms derive a part of their idiosyncratic palimpsests from recycled building blocks, becomes an important form of legitimation. And the fact that their practice of recycling is presented as humanistic makes them not only a prized reference, but also an example to emulate.[130]

Wang Shu and Lu Weniu's context-sensitive approach to recycling is also but one particular illustration of *architecture integrating notions of circular economy into its ability to innovate.* Ultimately, the Chinese architects bring an author's signature and a relatable form to ideas that otherwise could remain distant abstractions from the economic realm. Their "poetic recycling" gives a distinct architectural imprint to the notion that, in terms of everyday goods and services, we must transition from a model of linear production to a circular one — in which waste and by-product materials must be transformed and reintegrated through innovation, be it technological, behavioural or otherwise.

128. See Christele Harrouk, "On Recycled Architecture: 12 Proposals to Promote Adaptive Reuse," 15 Jul 2020. ArchDaily. Accessed 21 Jul 2020. <https://www.archdaily.com/931659/10-plus-proposals-to-promote-adaptive-reuse-and-introduce-transformative-ideas>

129. See Susanna Moreira. "Where to Apply Recycled Materials in Architecture and Urbanism?" [Onde usar materiais reciclados na arquitetura e urbanismo? 8 aplicações possíveis] 13 Jul 2020. ArchDaily. (Trans. Duduch, Tarsila) Accessed 21 Jul 2020. <https://www.archdaily.com/943256/where-to-apply-recycled-materials-in-architecture-and-urbanism-8-possible-applications>

130. Scarlett Miao. "Amateur Architecture Studio's Works on Contemporary Chinese Architecture with Recycled Materials" 06 Jul 2020. ArchDaily. Accessed 21 Jul 2020. <https://www.archdaily.com/942622/wang-shus-works-on-contemporary-chinese-architecture-with-recycled-materials>

Suddenly, circular thinking and circular design are seen as "shaping urban life." As increasingly addressed in media, from plastic waste upcycling to flexible urban grids that allow for renovation, this is seen as "a systemic shift that builds long-term resilience to provide environmental and societal benefits."[131] Architectural thinking is newly seen as contributing to an increased circularity of construction processes. Such inputs may translate into simply facilitating information on recyclable resources. Or they may correspond to the conception of more sophisticated frameworks by which architectural practice transitions into a full circular production model.

Inspired by the earlier experiments of Rural Studio with locally discarded industrial materials, Dutch collective Superuse concocted their name from the idea that the "use" of a given resource could be doubled, magnified and expanded, instead of simply abandoned after its first deployment. To put their concept to work, they pioneered an electronic exchange in which producers from a wider region could share their dismissed resources or leftovers, so that designers can then chart and integrate those flows of materials into their architectural concepts. They thus speak of "a continuous cycle of creation and recreation, use and reuse," which deploys context and availability so as to create "urban ecosystems from valuable residual flows."[132]

One of Superuse's necessarily diverse projects, a villa for a couple of art collectors completed in 2008, could then draw its steel profiles from a redundant textile machine located a few kilometres away from its site, while it's facade was "clad with wooden slats from redundant cable-reels harvested at a local cable factory." As the studio notably suggests, the waste materials "provided a continuous stream of new incentives to develop and refine the design." And, as it could be expected, "new shapes and innovative construction methods were needed to incorporate the found materials."[133]

While Superuse offers a practical avenue to immediately start internalizing circularity into design operations, other architects have engaged it as more of a systemic case study.

131. See Eric Baldwin, "Recycled Cities: How Circular Design Shapes Urban Life," 21 July 2020. ArchDaily. Assessed 21 July 2020. https://www.archdaily.com/944129/recycled-cities-how-circular-design-shapes-urban-life?

132. See About Us. Superuse-studios.com. Assessed 21 July 2020, https://www.superuse-studios.com/en/about-us/

133. See Erik Stekelenburg, Villa Welpeloo. Archello. Assessed 21 July 2020, https://archello.com/project/villa-welpeloo-2

A villa for a couple of art collectors by Superuse, 2008. Photo by Allard van der Hoek

More than a commons innovation through increasingly accessible technology, the attraction of circularity would then come from its business appeal for the industry.

In their 2016 book, *Building a Circular Future*, Danish architect Kasper Guldager Jensen, of 3XN Architects, joined consultant John Sommers, so as to demonstrate how circular economy principles applied to construction could be an important tool in "combating climate change." But, on the other hand, they essentially deliver a "positive business case" in which today's demolition costs could immediately double their value as earnings, while, due to resource scarcity, they could further increase in value over time.[134] In such case, as it has already been advanced in previous chapters, the model relies on creating a material passport, on specifically designing "for disassembly," and on ensuring biological and technical circularity for as long

134. See Kasper Guldager Jensen and John Sommers, *Building a Circular Future*, GXN Innovation, Copenhagen: 2016. Available for download at https://gxn.3xn.com/wp-content/uploads/sites/4/2018/09/Building-a-Circular-Future_3rd-Edition_Compressed_V2-1.pdf

as possible. As defended by the authors, implementation of their model of circularity at industrial building scale would allow for faster construction times, increased flexibility and optimized maintenance of building components.

While Superuse's innovative take on circularity is based on a small-scale, bottom-up and shared charting of resources, the other proposal rather relies on the building industry's need to remain competitive while responding to new social — and environmental — demands. As the authors implied, their proposals and case studies were really about expanding "the value of circular economy by introducing social and aesthetic concerns into the equation alongside material and construction costs."

This call for large-scale pathways based on a direct response from the construction industry is certainly necessary for a wider climax shift. Typically, these align with more general tendencies prompting initial efforts of decarbonization. There is nothing easier to start decarbonization than taking advantage of whatever technical efficiencies we can put our hands on, be it in energy production, mobility and in other sectors that are major carbon dioxide emitters. Logically, it is then in this direction that we must also look out for *architecture striving for every possible carbon-reducing efficiency in its construction processes, operation systems and material alternatives.*

We have already mentioned the emerging bio and green surrogates that are being readied to substitute widely used construction materials. The introduction of alternatives to concrete, as much as of timber or new biomaterials, will basically depend on newfound efficiencies. These will make them prone to reduce overall carbon footprints and even aim at negative emissions. As such, beyond the tech sector, startups in the building industry have already realised the potential magic of discovering the efficient building block that will revolutionize the construction market. So, from non-fired, construction-waste-made bricks[135] to mushroom-based bricks,[136] and from proficient 3D printing[137] to novel try-outs in biomimicry,[138] ideas have abounded — searching

135. See Rima Sabina Aouf, "Kenoteq launches brick made almost entirely of construction waste," 2 March 2020. Dezeen. Assessed 21 July 2020. https://www.dezeen.com/2020/03/02/kenoteq-k-briq-brick-construction-waste/

136. See Golda Arthur, "Making Houses out of Mushrooms," 30 August 2014. BBC. Assessed 21 July 2020. https://www.bbc.com/news/magazine-28712940

137. See Tom Ravenscroft, "ETH Zurich combines 3D printing and casting to make more efficient concrete structures," 2 July 2020. Dezeen. Assessed 21 July 2020. https://www.dezeen.com/2020/07/02/eth-zurich-3d-printing-concrete-casting-fast-complexity/

138. See Rima Sabina Aouf, "University of Stuttgart creates biomimetic pavilions based on sea urchins and beetle wings," 8 May 2019. Dezeen. Assessed 21 July 2020. https://www.dezeen.com/2019/05/08/university-stuttgart-biomimetic-pavilion-bundesgartenschau-horticultural-show/

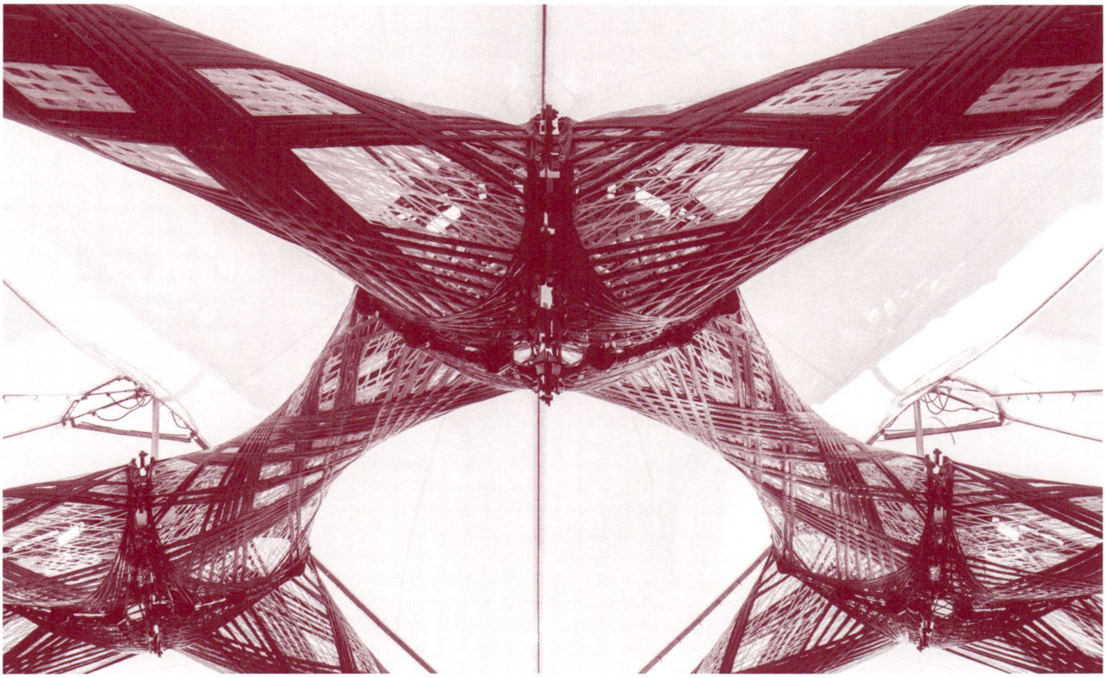

BUGA Fibre Pavillion
© ICD-ITKE

for the breakthrough or silver bullet that will change it all in terms of carbon reduction prospects.

Improved technological efficiency is also central to how imbedded building systems can contribute to reduce excessive carbon emissions. Building automation, for instance, relies on increasingly abundant and more precise sensors. And these allow for a closer monitoring of the energy flows that are being consumed or lost in the built environment. As such, both in the context of proposed Green New Deals and ongoing scientific research, some have defended it as a top political priority[138] — where there is actually the capability to impose it as a technological add-on to the adaptive reuse of existing structures. While the first to jump on board are the industries and corporations that are able to implement the technological fixes leading to smarter buildings,[140] one still must wonder if this is the most viable pathway for effective change, especially given the economic constraints in developing countries of the Global South.

139. See Casey Talon, "Intelligent building technologies are critical to a low-carbon future," 4 Oct 2019. Euractiv.com. Assessed 22 July 2020. https://www.euractiv.com/section/energy/opinion/intelligent-building-technologies-are-critical-to-a-low-carbon-future/

140. See Julie Seto, "How smart buildings can help combat climate change", 12 Dec 2018. Microsoft Azure. Assessed 22 July 2020, https://azure.microsoft.com/en-us/blog/how-smart-buildings-can-help-combat-climate-change/

Before we even get to full building automation, however, air conditioning and heating systems are still the biggest elephants sitting in our living rooms. These represent the most impactful bulk of carbon dioxide emissions from the building sector. Due to warming, air conditioning and cooling are inevitably growing, especially in developing countries. Not surprisingly, when it comes to greenhouse gases mitigation, these are also the ones that present the biggest potential for efficiency gains in the near future. So much so that, in Paul Hawken's *Drawdown*, alterations to the chemistry of our current refrigeration systems come in at an amazing #1 as the technological advancement that may imply the widest cutbacks of CO2 over the next three decades — that is, if we include "the management and destruction of refrigerants already in circulation."

Historically, air conditioning, like concrete or steel, can been seen as one of those critical inflexions by which architectural modernity became ecocidal. As one critic recently reiterated, forced air was certainly "one of the galvanizing moments in the development of architectural modernism."[141] One of its blunt effects, nonetheless, was that it drove architects away from early modern research into passive, climate-sensitive architecture, as Daniel A. Barber extensively accounts in his *Modern Architecture and Climate: Design Before Air Conditioning*.[142]

What we need now is a reversal in direction of such earlier history, but with the knowledge of more efficient carbon drawdown processes. If architectural design cannot perform an outright rejection of air conditioning, namely in favour of cleverer natural ventilation systems, it can at least quickly accommodate those gains in efficiency and carbon reductions that are promised for both this particular appliance and its architectural twin, heating.

Pathways related to technological efficiency in these arenas, as in any similar operative system that is central to the building industry, imply an urgent and permanent updating of the scientific and technical knowledge made available to the architectural field. This does not imply that every architect is to become a science or tech amateur, but rather that architectural

141. See Kate Wagner, "Forced Air: In Modern Architecture and Climate, climate control takes command," 9 July 2020. Architect's Newspaper. Assessed 22 July 2020, https://www.archpaper.com/2020/07/review-modern-architecture-and-climate-climaticization-takes-command/

142. See Daniel A. Barber, *Modern Architecture and Climate: Design Before Air Conditioning*. Princeton University Press, Princeton: 2020.

research increasingly relies on broader interdisciplinary allegiances. Both in academia and at the office, at the research lab or in tandem with major companies, as in architecture culture at large, incentives must certainly flourish in that direction. Because, as another chief pathway for a pressing paradigm shift, we must also see *architecture engaging in a wider range of collaborations towards multidisciplinary climate solutions for the global building environment.*

When it comes to carbon mitigation, we have already written here about the benefits that architects could derive from a more fruitful collaboration with landscape architects. And while research into sustainability already swarms in the architectural field, especially on its more progressive corporate side, interactions with ecological and climate science may also soon prove indispensable to create projects that are effectively environment-responsive. As cities are faced with increased environmental disruptions, complex problems in the urban realm necessarily require the allegiance of diverse technical skills and knowledge sets.

In a global context, however, other sorts of multidisciplinary collaborations arise as facilitators for architectural projects that, in given specific conditions, provide regional development with an eye on environmental issues. As seen in the example of South Africa's Langbos school, approaches to international cooperation can involve architects, manufacturers and the heirloom of previous architectural research. Such collaborative endeavours can allow for relevant technological leaps where climate efficiency gains are still a distant mirage. They also provide for a transfer of knowledge into particularly deprived contexts.[143] In spite of the danger of neo-colonial temptations, namely of a commercial nature, we must hope that this is one potential avenue to bring climate action to settings that, ultimately, are the most vulnerable to an encroaching ecological crisis.

While accounting for the growing majority of the planet's population, in economic terms, countries in the Global South are far from being able to address the oncoming environmental emergency on their own. And if this demands Green New Deals

143. See Pedro Gadanho, "Emergence vs. Emergency: How Knowledge Must Follow Fashion." In F. Le Roy, N. Wynants, D. Hoens and R Vanderbeeken (eds.), *Tickle Your Catastrophe, Imagining Catastrophe in Art, Architecture and Philosophy.* Academia Press, Gent: 2011.

MoMA: Uneven Growth exhibition by NLÉ

that are unimaginably different from those being pushed in Western countries, it also implies radical new collaborations that may anticipate and implement leapfrog solutions to eminent catastrophes in these countries' exploding built environment.

This is probably the reason why the brief encounter between Lagos-based Nlé Architects and a young architecture collective from Spain remains, for me, one of the most exciting collaborations born out of MoMA's curatorial project *Uneven Growth: Tactical Urbanisms for Expanding Megacities*. While the exhibition addressed the problem of increasing urban inequality, it inevitably intersected broader environmental issues. And this was also the case of the research partnership established between the two very distinct architectural studios that were engaged to look at the future of Lagos, in Nigeria. The temporary collaborative decided to focus on energy, water and transportation issues, and this pushed them to come up with inventive bottom-up solutions for a truly out-of-the-box vision of sustainability. Including MadMax-style revamped generators and redundant high-tension masts turned to renewable energy production, their project specifically aimed to reduce grid dependency, striving to imagine alternative sources of energy generation in a bustling megalopolis. On other fronts, it proposed new architectural forms to appropriate and regenerate the city's polluted water surfaces. Finally, it pushed for green alternatives in terms of mobility, imagining an urban scenario where biking and sustainable water transportation become dominant.

While underlining the creative potential of unexpected alliances between diversely oriented actors in the architecture world, this collaboration within the cultural realm of the museum also hinted at another essential, broader pathway for a more resilient future: *architectural thinking contributing to reconceive shared infrastructures and complex urban environments for a new environmental paradigm.*

Walkable and bikeable cities, collective green spaces, urban afforestation and farming, community clusters around recycling or composting, climate-resilient retrofits, microgrids and alternative energy generators in urban contexts, water and pollution management, flood and heat island effect mitigation designs, are but some of the metropolitan novelties that are arising from the need to address the several fronts of a global ecological crisis.

While much of these aspects draw on adaptation policies,[144] they also allow for novel design responses, as much as for the potential to envisage new architectural and urban typologies. With good reasons for municipalities to lead demands for climate action — they occupy only 2% of the planet's surface, but they consume nearly 80% of global energy and emit over 60% of greenhouse gases worldwide[145] — these are the first and foremost stages for architecture to contribute with exciting conceptual innovations tackling the ecological crisis.

And in fact, a new wave of speculative designs with global expression has often put architects on the front line of newly imagined city infrastructures, which directly respond to clamouring, often unanswered environmental concerns. As such, India-based Studio Symbiosis envisaged air purifying towers in

144. See, for example, *EU Urban Adaptation Policies*. Climate Adapt: Sharing Adaptation Information Across Europe. Assessed 23 July 2020. https://climate-adapt.eea.europa.eu/eu-adaptation-policy/sector-policies/urban

145. See Ryan Plummer, Darby McGrath, Jessica Blythe & Julia Baird. "Three ways cities can take the lead on climate change", 19 November 2019. Assessed 23 july 2020. Quartz. https://qz.com/1750042/three-ways-cities-can-take-the-lead-on-climate-change/

146. See Lizzie Crook, "Studio Symbiosis proposes Aũra towers to alleviate air pollution in Delhi," 9 January 2020. Dezeen. Assessed 23 July 2020, https://www.dezeen.com/2020/01/09/studio-symbiosis-aura-towers-air-pollution-delhi-architecture/

hyper-polluted New Delhi,[146] while Brooklyn-based Framlab invented tree-like modular vertical farms "to provide low-income neighbourhoods with access to fresh produce."[147] And South-African architect Nicole Moyo devised an acclaimed community-based waste-to-energy management hub for informal settlements,[148] while Giovanni Vaccarini Architetti's *Powerbarn* in an Italian city's industrial quarters showed that renewable energies sites can also be an attractive opportunity for experimental form-making.[149]

Aũra towers to alleviate air pollution in Delhi by Studio Symbiosis

There are also those architectural offices, such as London-based Superflux, which actually take advantage of cultural circuits to pursue a consistent line of speculative research on what has already been funkily dubbed as "climate futures." In consecutive exhibition installations, the collective has fictionalised the artefacts and behaviours of coming urban living, ranging from domestic hydroponic farming to de-digitalised tools for a new hunter-gatherer society. Making tangible an array of dystopian possibilities, they claim to use "narrative and speculation as a means of exploring

Glasir: tree-like modular vertical farms by Framlab

147. See Kristine Klein, "Framlab proposes modular vertical farms for Brooklyn neighbourhoods," 31 January 2020. Dezeen. Assessed 23 July 2020, https://www.dezeen.com/2020/01/31/framlab-modular-vertical-farms-brooklyn/

148. See Lizzie Crook, "Nicole Moyo's waste-to-energy system will "mobilise communities in informal settlements"," 1 April 2019. Dezeen. https://www.dezeen.com/2019/04/01/nicole-moyo-ukubutha-waste-energy-infrastructure/

149. See Shonquis Moreno, "Giovanni Vaccarini Architetti's Powerbarn Treads Lightly—While Making a Big Impact," 21 November 2019. Metropolis. Assessed 23 July 2020. https://www.metropolismag.com/architecture/giovanni-vaccarini-architettis-powerbarn-treads-lightly-while-making-a-big-impact/

150. See Drew Zeiba, "Superflux brings climate change home in a speculative Singapore apartment," 17 January 2020. The Architect's Newspaper. Assessed 23 July 2020. https://www.archpaper.com/2020/01/interior-superflux-climate-change-speculative-singapore-apartment/

complex problems that are often discussed in terms of data and abstract projections."[150] Intuitively, they actually point to another direction or pathway through which we can have *architecture creating climate crisis awareness out of its specific disciplinary tools.*

Such a pathway can naturally stretch from an active participation in more or less specialised circuits of cultural production, to a pedagogical responsibility that, as already mentioned here, can touch on clients, city authorities, media, and other social actors. Pedagogy starts at tables, in conversations with project promoters or over the drawing board, but it extends to many other possibilities. As one critic pointed out, awareness can be raised through project stipulations.

So, "even if architects alone can't reduce the embodied carbon of their projects," Gideon Fink Shapiro suggests, "they are the party most likely to have the client's ear" — and thus the ones that most probably can "affect the supply chain through their specifications."[151]

151. See Gideon Fink Shapiro, "Concrete, Steel, or Wood: Searching for Zero-Net-Carbon Structural Materials," 15 Jan 2020. architectmagazine.com. Accessed 18 May 2020, https://www.architectmagazine.com/technology/concrete-steel-or-wood-searching-for-zero-net-carbon-structural-materials_o

This power to influence opinions and decisions can emerge from the impact of individual speculative proposals, or, on the other side, from the complementary role that professional organizations in the architectural arena have in collecting information and formulating reports for the industry. Yet, curiously, when it comes to environmentally driven architecture, ideas competitions have consistently proven to be one of the best means to achieve multiple ends.

Major actors in the building industry have already realized the potential of ideas competitions. As such, they are deploying open calls so as to prove their sustainability commitments — with the LafargeHolcim Awards for Sustainable Construction, and its $2 million cash prizes for each three-year cycle, as a generous illustration.[152]

152. See LafargeHolcim-Foundation.org. Assessed 23 July 2020 https://www.lafargeholcim-foundation.org/awards/introduction

Similarly, new and existing international competitions in the architectural arena are increasingly directed towards responses to the climate crisis. This may happen by mandate and topic — as happens with the Land Art Generator's open calls under the motto "renewable energy can be beautiful",[153] or with the example of a recent Young Architects Competitions edition dedicated to plastic reuse, in "The Plastic Monument".[154] But, more interestingly, this tendency may also materialize because participants decide to make it a topic of their entries — as has increasingly been the case with the latest editions of the eVolo Skyscraper competition.[155]

153. See landartgenerator.org. Assessed 23 July 2020, https://landartgenerator.org.

154. See "Plastic Monument: An Architectural Call for our Planet" 08 Apr 2019. ArchDaily. Accessed 23 Jul 2020. <https://www.archdaily.com/914526/plastic-monument-an-architectural-call-for-our-planet>

155. See Mackenzie Goldberg, "Here are the creative skyscraper designs selected as winners of eVolo's 2019 Skyscraper Competition," 30 April 2019. Archinect. Assessed 23 July 2020, https://archinect.com/news/article/150134253/here-are-the-creative-skyscraper-designs-selected-as-winners-of-evolo-s-2019-skyscraper-competition

In any case, competitions provide dissemination for the issue of the environmental emergency, but, in a sort of minor climax, also a sudden wealth of new concepts to be shared and emulated by the architecture community.

Speaking the visual language of speculative architectural proposals, some competitions end up promoting ecological-

Ukubutha by Nicole Moyo

conscious narratives through the specific abilities of form-making. As much as students in architectural schools may respond to traditional briefs with emerging environmental concerns, calls for ideas — as, in other contexts, curatorial strategies — are a fundamental way of making architectural culture progress towards new orientations and sensibilities.

In parallel, as amply recorded here, some specialised media have followed on the footsteps of the most progressive press around the globe, and have played a fundamental role in disseminating both environmental-prone contents and critical reflections on the need for sustainable change. Reporting abundantly on the consequences and implications of the climate crisis, some architectural media have assumed their responsibility in making the topic real for wider audiences.

With amplified online visibility, the environmental emergency can indeed be said to have become viral — and this is certainly a necessary contribution to both raising wider awareness and slowly changing architects' sensibility towards an unavoidable

Powerbarn by Giovanni Vaccarini Architetti

climax change. As self-aware, online platforms have pointed out in detail — also speaking from or to locations in the Global South[156] — the year 2019, in particular, was a turning point towards the inevitability of such realization in the world of architecture.

With such dissemination efforts already under way, the next implicit move would be to progress from the growing pains of awareness to the actual shift in practice that will see *architecture embracing ecological regeneration, interspecies design and environmental mindfulness as irrefutable tenets underlying any aspects of its future production.*

Repair, restoration and regeneration are the new seminal R words that, with the eminence of ecological crisis, have come to join the already commonly accepted musts of reduce, reuse and recycle. If the latter were a behavioural mantra directed at the human use of material flows, the more recent requisites are

156. See Niall Walsh. "Como a arquitetura respondeu às mudanças climáticas em 2019" [How Architecture Responded to Climate Change in 2019] 09 Jan 2020. ArchDaily Brasil. (Trad. Sbeghen Ghisleni, Camila) Assessed 23 Jul 2020. <https://www.archdaily.com.br/br/931447/como-a-arquitetura-respondeu-as-mudancas-climaticas-em-2019>

redirected at how humans actually engage with Nature. This is a subtle but important shift. It should represent a game changer for architectural practice as well.

In *The Collapse of Western Civilization: A View from the Future*, science historians Naomi Oreskes and Erik M. Conway trace a vivid account of how our present trajectory could look to a future historian one hundred years ahead. In their intriguing science fiction narrative, while people seemed to understand the hard facts of a potential climate catastrophe, they bluntly failed to act on it. As the authors put it, the startling part was that "knowledge did not translate into power."[157]

157. See Naomi Oreskes and Erik M. Conway, *The Collapse of Western Civilization: A View from the Future*, Columbia University Press, New York: 2014.

In other words, there is a real risk that we fail to use any available tool to properly reengage with the planet's ecosystem, and strive to invert our current crash course. As a cautionary tale, this warns us that if political power fails, if scientific clout wanes, if fossil fuel lobbies prevail, we should eventually turn to any form of everyday practice that offers some glimpse of potential change — from personal behaviours to policy-making, from activism to industrial reconversion, with contributions from both the collective and the private sector.

Individual architects regularly claim that they are powerless in the face of dire collective challenges, such as those posed by the climate crisis. As some authors have noted, this suggests disillusionment, helplessness and, I would add, sometimes even a discrete cynicism. In contrast, architects would need to "acknowledge impact and responsibility," as much as to "reframe the extent of architectural influence."[158] Also, rather than having sustainability as the ultimate goal, that degraded notion should be only one "non-negotiable" baseline on which we build more ambitious and balanced connexions with surrounding ecosystems. Disciplines that plan and mould the built environment, we must assume, can adopt and impose logics of action that may astutely alter the dangerous game we have been playing with our environment.

158. See James Woodall. "Reframing Climate Change as a Local Problem of Global Proportion: 4 Ways Architects Can Deliver Change." 28 January 2019. Assessed 24 July 2020. https://www.archdaily.com/909776/reframing-climate-change-as-a-l...roblem-of-global-proportion-4-ways-architects-can-deliver-change

While painting a bleak scenario for the approaching environmental devastation, architect and writer Richard Buday recently reminded architects that they can be pretty persuasive and are still the ones from which "the construction community" take their cues. In offering prompts for the near future, perhaps we should aim at more than the immediate adaptation to catastrophe that Buday backs. Even so, this author was on target when he points that it is about time architects "begin accepting Mother Nature's terms of surrender."[159] After missing many opportunities to move towards a carbon-net-zero built environment over the last decades, what little power of persuasion that we have left should make us start moving towards truly ecologically friendly practices.

159. See Richard Buday, "What Architects Don't Get About Climate Change" 9 March 2019. Common Edge. Assessed 24 July 2020. https://commonedge.org/what-architects-dont-get-about-climate-change/

160. See Darran Anderson, "As environmental catastrophe unfolds, we need architecture that is more than just green," 23 January 2019. Assessed 24 July 2020. https://www.dezeen.com/2019/01/23/darran-anderson-opinion-modular-adaptive-architecture-environment-anthropocene/

As we have described here, intentions in that direction first flourished when architecture organizations and leading professionals in the field declared climate and biodiversity emergencies. However, as we also recognised at the early stages of this narrative, to start changes in effective practice seems less obvious. If vows are easy, effective commitments are harder. Nonetheless, the call is ringing insistently — and it is not only about the urgency of decarbonization.

As another author urged in early 2019, with the environmental catastrophe lurking around us, "we need architecture that is more than just green." For Darran Anderson, architects would need modular and adaptive innovation, which "anticipates and responds to the changing environment." Instead of gearing the trend of smart cities towards "surveillance or sociological efficiencies," he suggests, "data-driven technology could be used to drive and support urban ecosystems, minimising waste and energy and maximising growth."[160] Although in that arena greenwashing always lurks around some corner, this is just one technological route to achieve the ecological turn we have already hinted at, namely in the "Thinking Like a Building" chapter.

Heading towards an ecological design practice, but considering some of the techniques involved in biological carbon removal and ecosystem restoration, we can say that architecture is still far from integrating concepts such as mineral weathering or rewilding as its own potential tools. While one involves the extensive use of certain minerals for their ability to store carbon, on the other side of the spectrum, the other means creating the conditions for nature to again take over sites that have been polluted or depleted. Far as they may seem from your usual architecture practice, the picture starts to change when one realizes that broken down concrete is itself a mix with carbon storage potential. This means that, beyond the simple banishing of new constructions in virgin land, demolitions of existing structures themselves, if well-conceived, could become instruments of ecological regeneration. This just implies that we combine a little more knowledge with our usual creative abilities. And the same goes for much current scientific research that, in one way or another, has the potential to penetrate how we conceive our built environment.

For sure, as we have noted, we will have to design with Nature, not against it. As recently stated in a book advocating this position, we must improve "ecosystems by contributing to ecosystem services rather than depleting them."[161] And, while much investment is being made in climate science, research units and labs around the globe are busy creating the base knowledge from which many built environment applications and design ideas will have to emerge. Again, this suggests interdisciplinary collaborations must necessarily spread. And it also implies that, as again already affirmed here, we must design with more than human subjects in mind.

161. See Frederick R. Steiner, in Frederick R. Steiner, Richard Weller, Karen M'Closkey, and Billy Fleming (eds.) *Design With Nature Now*, Lincoln Institute of Land Policy, Cambridge, Massachusetts: 2019

In this sense, it is fortunate that interspecies design is finally moving away from academic, experimental or cultural niches, and is becoming a notion that your local architecture practice now wants to proudly deploy. Or so it seems, when we read in another architectural newsletter that Pittsburgh-based Epiphyte Lab, besides operating cross-disciplinarily "between design and science," also declare that "architecture is fundamentally a part of planetary ecology and thus has to operate with empathy to all species." Thus, their research includes concepts such as the retrofit of old Eastern European blocks with a new façade that will welcome both local flora and fauna, while also housing "camouflaged energy-harvesting devices."

This is hopefully only the beginning of widespread ecologically driven practices, which in a few years we will assume as architecture's "new normal." Biotecture, biomimicry and other biophilic design strategies should naturally become the common ground from which exciting architectural styles and paradigms should bloom.

Biomimicry, for instance, has been fiercely championed by British architect Michael Pawlyn as a key to new forms of regenerative design. As the translation of "functions and processes found in biology into solutions that meet human needs" — and opposed to the deceptions of sustainability — biomimicry would be instrumental to "the development of restorative and renewable systems where output is always greater than input, beneficial for humans and other species." In this way, his Exploration Architecture studio seeks to learn from exoskeletons or even entire ecosystems, so as to find novel energy and material efficiencies — as demonstrated in his Sahara Forest Project, in Qatar, where "a seawater-cooled greenhouse" is "modelled on the Namibian fog-basking beetle, an insect that harvests its own fresh water in a desert." [162]

Likewise, one of *Drawdown*'s "coming attractions" for decarbonization, the notion of "living buildings," might act as more than just a certification process and an important inspiration for new design imperatives — even if these include motivating principles such as "net positive" energy, water and waste; the ability of a construction to grow its own food; or "habitat exchange," in which for every developed acre for human use, one is set for a protected natural habitat.[163]

When it comes to the future of our built environment, ecological regeneration should become architecture's next unswerving obsession. And following on such priority, we could finally be drawn to consider a less-discussed option for environmentally inclined architectural practices. This is the pathway in which we would hopefully foresee *architecture partaking in the conceptualization and implementation of new ecological commons.*

162. See Lizzie Crook, ""We fooled ourselves that sustainability was getting us where we needed to go" says Michael Pawlyn of Architects Declare." 7 Oct 2019. Dezeen. Assessed 24 July 2020. https://www.dezeen.com/2019/10/07/michael-pawlyn-architects-declare-interview-regenerative-architecture/

163. See Paul Hawken, ibidem.

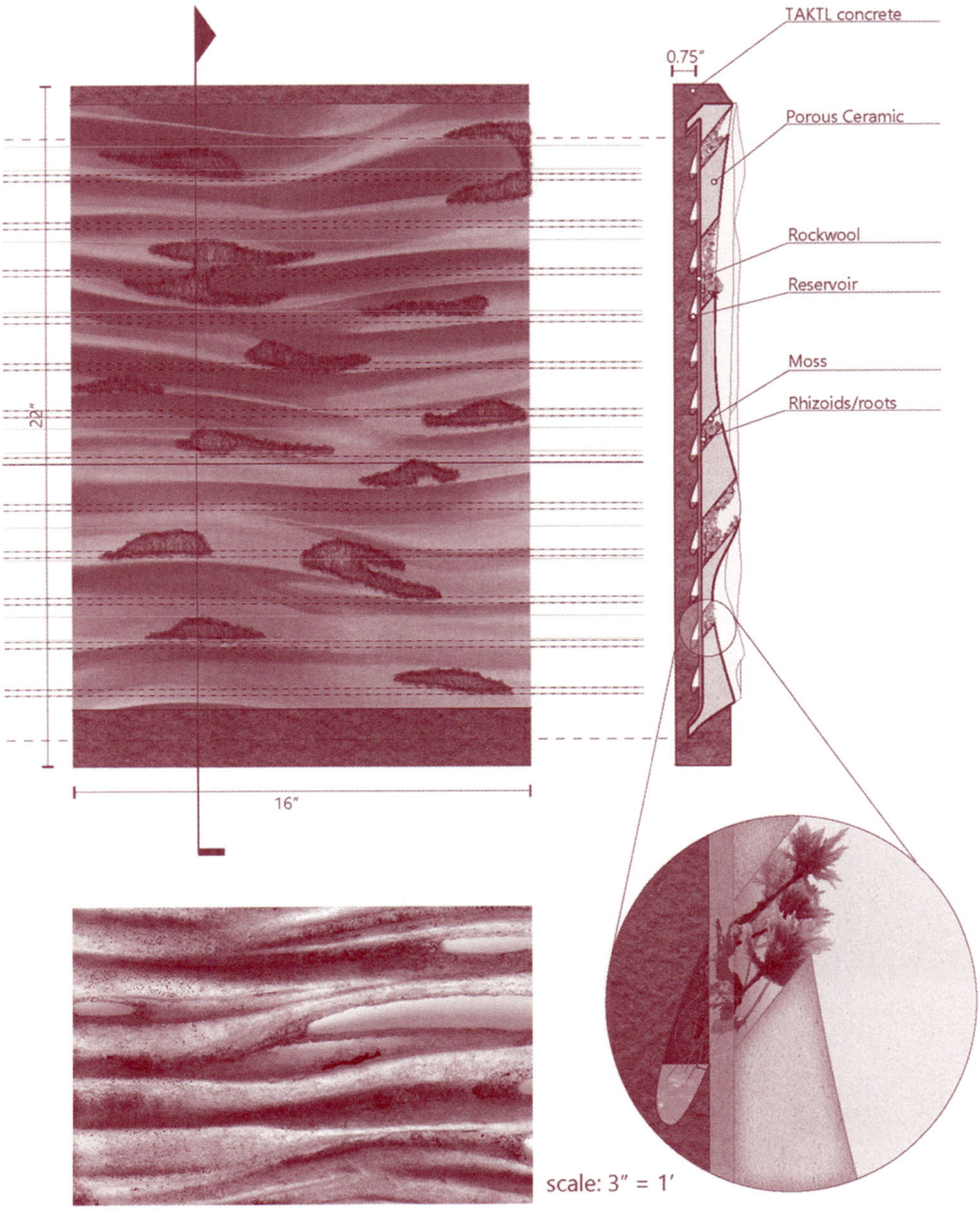

Embedded Biomass in Porous Ceramics, Acadia 2018 Recalibration: on Imprecision and Infidelity: 38th Annual Project Catalog, 2018. © Dana Cupkova @Epiphyte Lab: Moss Regimes

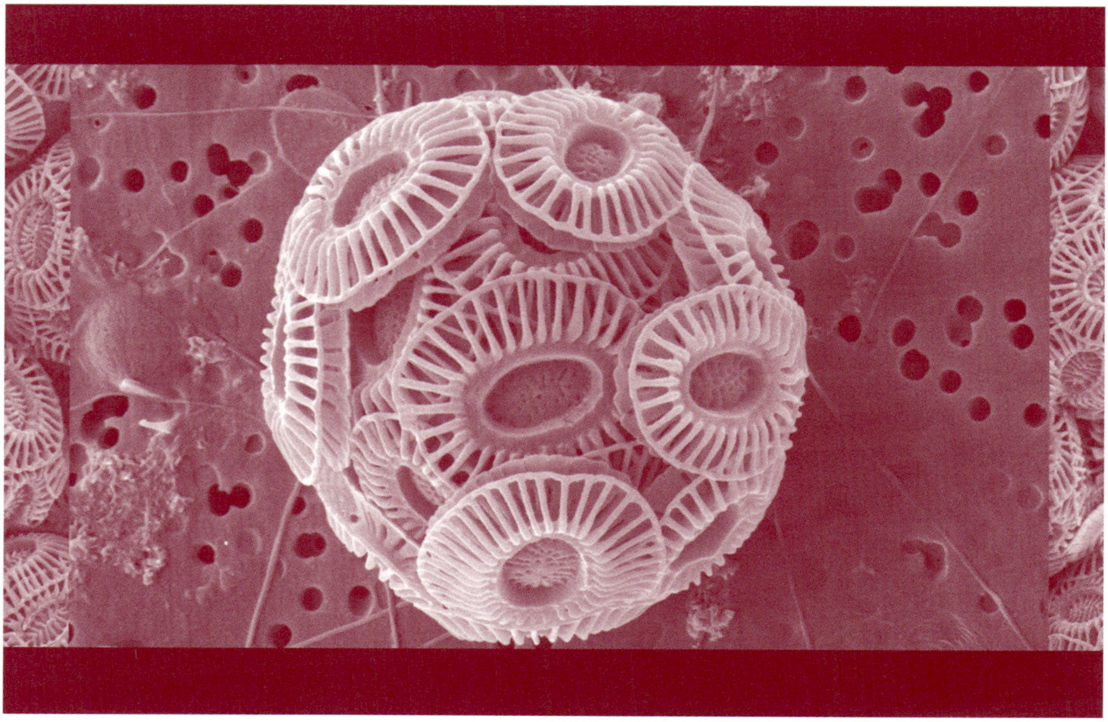

The image is an amalgam of creative commons images from Wikipedia under the search coccolithophore by Exploration Architecture

While the call for a return to the commons is often eloquently put in terms of the need to fix shattered social contracts, the comeback of such anthropocentric notion may come sooner — and in a different guise — than is usually expected. The renewed character of the commons can actually emerge due to the economic effects of partial, unanticipated environmental emergencies, which the 2019 coronavirus outbreak only made strangely tangible.

Rather than financial stress for the 99%, as it was perceived after the 2008 Great Recession, soon it may be case that the small everyday climate catastrophes are the ones that will drive us to reconsider the pertinence of the commons as a mutual, ecologically driven support system. Among predictable food crises and thorn welfare States, we need alternatives for resilience that simultaneously offer an alternative political response and a more adequate ecological balance.

In an essay called "Enterprises of Survival," a group of mostly Eastern European architects, including the well-known Stealth collective, recently pointed in that direction. As others before, they suggested that the dispossessed creative class may be the one that actually engages in re-enacting the urban spaces that are abandoned in the aftermath of environmental disasters — as happened with initiatives of urban farming or collective living in economic-disaster-stricken cities.

As they put it, this means a willingness to "construct collective futures on the ruins of a broken system." And such experiments are already popping up, necessarily happening in small-scale communities such as Rotterdam's 2009 *City in the Making*.[164]

164. See Ana Džokić, Marc Neelen, Jere Kuzmanić, and Predrag Milić, "Enterprises of Survival". 5 Dec 2019. E-flux. Assessed 22 July 2020. https://www.e-flux.com/architecture/collectivity/303729/enterprises-of-survival/

According to these architects, such "training grounds" for community ownership give us a hint of guidance for the organization of other, scaled-up social models. But, most importantly for the climax change we advocate here, those experiments also point to two important lessons. First, any revamped notion of commons now implies ecological self-sufficiency and the overall care for a surrounding ecosystem in which humans are only one of the participants. Second, these new ecological commons ask for alternative modes of architectural practice.

From how the architects involved in "Enterprises of Survival" described their initiatives — as promoting "horizontal collaborations, open protocols, and forms of participation in cooperatives" — it became clear that the way they produced architecture stood "in stark contrast with the armies of interns, vertical hierarchies, or short-lived ambitions that characterize most of the profession."

Vis-à-vis the diverse social organizations that may arise out of a widespread environmental crisis, we also do need distinctive visions for how architectural practice may unfurl. Tellingly, the activities they described forced the intervening architects "to delve into unexpected fields of knowledge, to become ad-hoc finance experts, autonomous energy systems pioneers, self-styled 'business developers,' and more."

Stad in de Maak,
CC BY-SA 4.0

In similar fashion, the pathways that we have just tried to catalogue here speak of a paradigm shift in the architectural profession's guiding principles, as much as an expansion of the traditional limits of architecture practice. Long as this inventory may already be, the examples reported here do not in any way exhaust the extensive range of practice possibilities that are now available to architects — if they want to amass solutions and design ideas towards the *Great Transition*.

Any novel combinations of the pathways described above, as small-scale as they may be, will add to a necessary incremental change. Each little step will lead to other steps, each building upon its precedent. Which means that an historic transformation of architectural practice in the face of the ecological emergency is not out of reach — or that far off on the horizon. Pathways are available. It really is a question of how soon we set foot on those paths.

10. The Time is Now: Welcome to the Anthropocene

As awareness of a problem grows, so too concerns and worries take over the human mind. It is one of the species' basic survival mechanisms. As clouds mount on the horizon, we start buzzing about getting ready for the storm. As danger is sensed, we move into alert mode.

However, if danger is not instantaneous, as a species we have also developed ways to protect our psyches from unwanted stress. We have a great ability to postpone any fear that does not represent an immediate threat. Other, everyday anxieties will take precedence. If the sense of urgency is not absolute, this psychological mechanism will prevent us from taking apparently unnecessary action.

As social researcher Elke Weber points out, "when the challenge seems overwhelming and existing solutions inadequate, there is good reason to turn away from the problem." And as the climate change debate has largely demonstrated, we may prefer to focus on the uncertainties of existing predictions, and "use them as an excuse to delay action." With uncertainties ranging to humanity's technological and political ability to respond to climate hazards, comes the temptation to avoid or adjourn necessary behavioural changes. With the insight that we need changes across "social systems, economic systems, energy technology, personal and collective energy use," as well as our "consumption habits," comes a sense of impotence.[165] Lack of control over such a large problem may immobilize individuals, as well as whole countries. We may be lured to wait and see. In the case of the resulting environmental emergency, however, such inaction can turn out to be catastrophic.

165. See Elke U. Weber, "Climate Change Demands Behavioural Change: What Are the Challenges?" Social Research; New York Vol. 82, Iss. 3, (Fall 2015): 561-580,861.

If the effects of the ecological crisis are perceived as still far off on the horizon, we will find a way to defer any potential solutions. Economic, political, social or psychological reasons will provide the excuses to maintain a comfortable *status quo* while it may last. Despite all warnings, it is difficult to grasp that, in the case of an environmental breakdown, a late response may not be an answer at all.

If no adaptation or transition is planned ahead, by the time it sets in, a late reaction will only be slightly mitigating the lasting disasters of turning successive climate tipping points — from carbon dioxide levels to permafrost release of methane, from warming temperatures to unstable storm systems, from melting glaciers to rising sea waters, and from ecosystem collapses to full mass extinctions. Once these processes gain momentum, scientists point, their impacts will be felt for hundreds of years. Beyond the moment feedback loops lock in, any human response — as technologically advanced or as risky as it may be — will only very slowly affect any resemblance of a return to the ecological stability that made humans thrive.

Thinking and acting in terms of geological time, as we are asked to do in the case of the climate crisis, is intrinsically difficult. We are told that we are fast accelerating into this new geological period characterised by human intervention on the planet's surface. The Anthropocene, as they call it, is an understandable notion. And yet its time scales are hard to fathom, even if just because we are going through it — unlike other geological periods, which we observe from afar as well-defined historical epochs.

For sure, we can still understand that, depending on our scientific choices, the beginning of the Anthropocene can be located 70, 250, 500 or 15,000 years ago. Either if we favour an early beginning with the Agricultural Revolution, if we cling to Globalisation or the Industrial Revolution, or if we favour the more radical take of a Consumer Revolution triggering it after World War II, we can still relate to the onset of this new geological era in terms of human history. Yet, it may be trickier to grasp the extent of the Anthropocene, and the amplitude of its outcomes, going into the future.

Given the usual time scope of a geological era, it is also difficult to apprehend that, as much as humans have geared such a slow shift into full pace, they may also start to correct its course. We now understand how we shaped the Anthropocene into being. We also have the knowledge of how to make it somewhat less of a noxious episode in the long history of the planet.

Even if stalled by aspects of our current economic and political systems, we apparently have the scientific and technological capability across diverse fields of action to conceive a major ecological regeneration of our built environment. This is perhaps the best reason why we should fully embrace the notion of a human-moulded planetary era. And, yet, this does not necessarily mean we must embark on tricky and uncertain geoengineering practices, such as spraying more polluting aerosols into the troposphere.

As Braden Allenby has pictured it, focusing ourselves exclusively on "solving" particular climate change issues would correspond to treating symptoms, rather than "attacking the real disease" — which this specialist in sustainable engineering equates with the reality of the Anthropocene itself. While Allenby sees climate change as a management issue involving adaptation and mitigation, he rightly underlines how reducing the atmosphere's carbon content to pre-industrial era levels would still not restore the planet to that period.

"Complex adaptive systems do not have a default setting to which they can revert," Allenby writes. So, while we shoot silver bullets at isolated aspects of the climate crisis, we worryingly reveal "the inadequacy of our nascent efforts to respond to the challenges of the Anthropocene." Basically, the reductionist approach of addressing one discrete problem at a time is useless when dealing with the "emergent behaviours generated by seven billion people with their vast array of institutions, cultures, and economic and technological systems." As Allenby highlights, we simply cannot "de-Anthropocene the planet."[166]

166. See Allenby, Braden. Climate Redux: Welcome to the Anthropocene. Issues in Science and Technology; Washington Vol. 31, Iss. 3, (Spring 2015): 37-39.

But perhaps we can work through this inevitable condition to generate some new planetary balance. If human activity now rivals and conditions the "natural systems that govern the planet's habitability," Allenby points, then humanity must now "develop the capacity to manage these complex, interwoven systems." Only this "integrated planetary perspective" would shed some light at the end of the climate emergency tunnel. Instead of aiming at reversing human impacts on natural systems, we should accept human intelligence and its tools as evolutionary capabilities with which to eventually achieve the "sustainable design and management of a planetary ecosystem that includes the human system." As Allenby concludes, what is needed now "is the courage to perceive and accept the world as it is today, and to appreciate the difficulty of responsibly managing it."

Rather than just feeling ironically welcomed to the Anthropocene, we should indeed welcome it and take full acknowledgement of its implications. Further, if by triggering it we have driven ourselves to the current state of disarray, perhaps we can also design our way out of it.

This actually became the tenet of a curatorial project I had the opportunity to launch, which definitely allowed me to dive into issues of climate change and the environmental emergency. As underlined in its subtitle, the *Eco-Visionaries, Art and Architecture After the Anthropocene* exhibition assumed that we are already working from within, and after, the condition of the Anthropocene. It presupposed the Anthropocene as a given, but it also posited that its understanding gave us new tools to envision alternative futures beyond an inevitable ecological catastrophe.

Many of the more than 70 participants in the exhibition's multiple venues offered critical views on the critical moment we are living. Others, however, including those from the architecture and design fields, explored scenarios or revealed emergent practices in which we could foresee solutions and ideas like the ones discussed in this book. Artists and architects included in the show and accompanying publication had already fully welcomed the Anthropocene condition — and were actively and critically concocting its next steps.

Curiously, from its initial iteration in Lisbon's Museum of Art Architecture and Technology, in the Spring of 2018, to its version at the Royal Academy of the Arts, in London, by late 2019, reality quickly caught up with the exhibition's intent to further the discussion on how the design disciplines could contribute towards climate action. If in Lisbon the exhibition's

Eco-Visionaries - MAAT - 2019 © Bruno Lopes/Maat

contents were greeted with puzzlement and apprehension, less than two years later the smaller collection of proposals brought to London was already deemed as lacking and insufficient. The lesser number of projects may have been misleading, but, more importantly, over that short period awareness and activism around the climate emergency had both bloomed, prompting the demand for more immediate and impactful action. Only a couple of years after it appeared, the *Eco-Visionaries* project was already outdated and obsolete. And this was positive.

One of the most dramatic features of the Anthropocene is precisely its acceleration of geological time, bringing planetary systems dangerously close to human time-scales. Any climate graph you see these days bears that infamous hockey stick shape that brings every curve to an impossible slope over the last half century.[167] All those curves announcing heightened increases in carbon dioxide,

167. See Chris Mooney, The Hockey Stick: The Most Controversial Chart in Science, Explained, 10 May 2013. The Atlantic. Assessed 11 August 2020. https://www.theatlantic.com/technology/archive/2013/05/the-hockey-stick-the-most-controversial-chart-in-science-explained/275753/

Eco-Visionaries - Royal Academy London - 2019

temperature and whatever other indicators scientists are looking at, tell us only one thing: the time is up for drastic change.

As some desperately warn, we may still opt between the alternatives of either letting "business as usual" drive us to environmental catastrophe, or immediately start the changes that are required to mitigate the worst consequences of an unfurling climate crisis. But, surely, now is the time for that option to be enacted. And that active decision becomes a matter of individual responsibility across a horizon of collective choices. The change may start at the personal level, but it is only significant when it accrues into collective action on every front — from the diverse professional and disciplinary fields to wider behavioural shifts, from the private sector to governmental policy-making, from discrete rural locations to the scale of major urban settlements, and from communities to the realm of democratic decisions beyond national States.

We have welcomed ourselves to the Anthropocene. So, now we are facing immense changes affecting an expanded notion of the environment. The global ecological system can no longer be contained in a traditional vision of a distant, external Nature. Because we have the planet as an interrelated built environment, as an interconnected built ecology, the planning and design disciplines bear a special obligation. These disciplines, where architecture is included, should be at the forefront of the retrofit operations of the *Great Transition*, but also of new hybrid conceptions of ecological regeneration in a growingly urban context. Thus, they should also be among the first to join the calls that "the time is now."

Inevitably, in a time of Anthropocenic acceleration, calls for timely action have only become louder. As Time magazine put it, if the turn to the 2020s seemed like the time "when an unknown virus spun out of control," we may instead look back at it as the period in which "we decided to keep driving off the climate cliff — or to take the last exit." Increasingly, the realization spreads that this is "a defining moment in the fight against climate change" — and that "what we do now will define the fate of the planet — and human life on it — for decades."[168] As other reputed, conservative media promptly claimed, it now was time to "flatten the climate curve."[169] For official voices in the European Union, the pandemic disruption — and its unexpected ability to both change our everyday behaviors and put a brake on our frenzied acceleration — could actually be seen as *the* opportunity "to rebuild better."[170] Before Covid-19, all the Green New Deals being pushed around the globe were mostly aspirational; after the pandemic, they became a must.

The coronavirus pandemic became a paradoxical dress rehearsal for the impacts of a palpable climate emergency, racing us to a crucial crossroads. The billions and trillions required to trigger a "recovery" from the biggest economic crisis since the Great Depression could be deployed to get us back to an environmentally destructive "business as usual." Or, as pledged in the European

168. See Justin Worland, "2020 is our last, best chance to Save the Planet", July 20, 2020. Time Magazine. Assessed 8 August 2020. https://time.com/5864692/climate-change-defining-moment/

169. See Leader Section, "A New Opportunity to Tackle Climate Change: Countries should Seize the Moment to Flatten the Climate Curve," 21 May 2020. The Economist. Assessed 14 May 2020. https://www.economist.com/leaders/2020/05/21/countries-should-seize-the-moment-to-flatten-the-climate-curve

170. See EU Science Hub, "Climate change: Now is the Moment to Rebuild Better." 14 May 2020. European Commission. Assessed 8 August 2020. https://ec.europa.eu/jrc/en/news/climate-change-now-moment-rebuild-better

Union, they could alternatively put us on track to effectively implement those Green New Deals that were already under discussion. As I wrote elsewhere, "despite all the death and pain it caused, the pandemic would turn into a paradoxical opportunity if only it would allow for a speedier reconception of what "the new normal" could look like — be it in terms of social inequality, racial disparities, resource extraction, a poisonous built environment or all the ongoing ecological disruptions."[171]

171. See Pedro Gadanho, "Pandemic Paradoxes: World-Making at its Crossroads" 2 July 2020. Transfer-Arch.com. Assessed 9 August 2020. https://www.transfer-arch.com/transfer-next/pandemic-paradoxes/

Under the economic retraction provoked by the Covid-19 lockdowns, the world experienced sudden and brutal degrowth, although certainly not of the planned kind we have hinted at here. Degrowth became the subject of both caricature and overstated hopes. But, as one writer asserted at British newspaper *The Guardian*, repudiating degrowth based on the experience of the pandemic would be "premature and unjustified." Such rejection deliberately confused the harsh pandemic experience of "the unplanned, abrupt cessation of vast swaths of economic and social activity" with a "thoughtful, democratic, managed and equitable downsizing of the economy," which advocates of degrowth cautiously defend.[172]

172. See Lola Seaton, "In the midst of an economic crisis, can 'degrowth' provide an answer?" 24 April 2020. The Guardian. Assessed 10 August 2020. https://www.theguardian.com/commentisfree/2020/apr/24/economic-crisis-degrowth-green-new-deal

Surely, the painful arrest in economic activity unveiled the immediate effects of an abrupt and global lack of growth — much like the environmental crisis can one day make it recurring. With reduced mobility and grounded airplanes, carbon dioxide emissions certainly abated. Yet, as many warned, the pandemic provoked only a short-lived and illusory deflection on an ongoing trajectory towards climate catastrophe. Unless the opportunity for a greener upturn was grabbed, ensuing economic recovery efforts could just make things bitter.

Bill Gates, who had notoriously predicted the pandemic, joined the chorus in mid-2020 to stress that if the coronavirus was awful, "climate change could be worse." Displaying large amounts of data, the billionaire philanthropist too was keen to point out that, even as we recovered from the virus, we would also need to "act now to avoid a climate disaster." In his view, the greatest pandemic lesson was how it revealed us "the consequences of inaction," especially in view of the next, inevitable global crisis. If the first measures he suggested were to let science lead the way — but also to make sure that solutions are effective for poor countries — the most poignant one was certainly that we must "start now." Because if a vaccine could quickly be found for the corona virus, in regards to climate change "it will take decades to develop and deploy all the clean-energy inventions we need."[173]

173. See Bill Gates, "COVID-19 is awful. Climate change could be worse." 4 August 2020. GatesNotes. Assessed 10 August 2020. https://www.gatesnotes.com/Energy/Climate-and-COVID-19

174. See Bill Gates, "Buildings are bad for the planet," 28 Oct 2019. GatesNotes. Assessed 10 August 2020. https://www.gatesnotes.com/Energy/Buildings-are-good-for-people-and-bad-for-the-climate

175. See Marcus Fairs, "Coronavirus money could solve climate crisis but hopes of change are "wishful thinking" says Rem Koolhaas," 11 July 2020. Dezeen. Assessed 11 August 2020. https://www.dezeen.com/2020/07/11/coronavirus-change-environment-rem-koolhaas/

Considering that one of Gates' previous posts was on how "buildings are bad for the climate" — and on "what we can do about it"[174]— we can readily understand how the plea to seize the moment of a greener rebuilt should particularly resonate in the architecture world. And if someone like Rem Koolhaas rightly warned us of the "enormous pressure" for a post-pandemic return to business-as-usual,[175] we must still reiterate that, beyond awareness, individual commitments and "wishful thinking," a swift turn of collective will is crucial for the architectural profession as a whole. From its organizations and its education systems to its everyday practice, there is an urgency to change the fundamental impulses that drive the discipline. New enticements are needed for architects to adopt the design behaviours that will respond to the climate emergency — and allow others to seize a definitive turn to the *Great Transition*.

Behavioural specialists now posit that, the environmental crisis being so tough to crack, when we find the ways to motivate people to adopt environmental action and more sustainable behaviours, we could virtually solve any other of the major challenges impending upon us today — including social and economic inequalities of global dimension. Such is the scale of the matter. And yet, in another lesson not be forgotten, the Covid-19 pandemic also revealed how behaviours can drastically change overnight. Many took the corona virus as an undesirable distraction from the momentum the climate debate was gaining. Instead, the virus provided only more clues on how we could seize the urgency of an environmental transition.

The behavioural scientist quoted at the beginning of this chapter argues that, when we are considering action and pattern shifts in face of the climate crisis "wicked problem," it is "imperative to consider and use the full range of human motivations and goals." This is one good reason why we must talk about climax shifts in the architecture realm. Desires, aspirations, ambitions, trends, fashion turns — as much as fear, embarrassment or discredit — can all justifiably be summoned when it comes to push this collective into the necessary path of action.

As Weber adds, "behaviour change can be facilitated by innovative no-choice defaults, changes in the *status quo*, or the rephrasing and reframing of choice option labels." We need both a "better appreciation of human motivation" and "a focus on the positive consequences of change." These same strategies are thus required to transform mentalities in the design realm. As much as common citizens, consumers and even policy-makers, architects must be convinced to overcome a sense of impotence and to start enacting those "initiatives that, in combination and multiplied by millions of potential initiators, would live up to the [climate] challenge."[176] We need the thousand pathways we hinted at in the previous chapter, but also the behavioural revolution that will make us fully embrace them.

176. See Elke U. Weber, ibidem.

With no silver bullets and no technological panaceas immediately available, with no single solution for a quick decarbonization or energy transition, so too comes the urgency to immediately start a collective change in architecture's professional behaviours and everyday design practices. Among many directions we have discussed here, this also implies that disciplinary hawks abandon their typical disdain of any issues that escape a self-centred formal discourse — and they too understand their responsibility in shifting the discipline towards an inevitable ecological commitment. Amid discussions of architects' social responsibility towards today's inequities and threats, but also amid a waning emphasis on marginal subjects of inspiration from other fields, this is about the major disciplinary shift that we need to theorize right now.

It may be true that, as one critic recently mused, the "end of history" and "the omnipresence of capitalism as the most powerful ideological force" has lead architects to accept an alleged inability to respond to emerging social challenges. Instead, by way of what Kenneth Frampton has deemed an endemic "neoliberal indifference," they took refuge in either "forging links with the world of economics and politics" or "retreating into the realm of phenomenology."[177] Contrary to the emancipatory spirit of earlier architectural discourses, this corresponded to a de-politicisation and a dismissal of responsibility in both practice and theory, which the global environmental crisis now makes untenable.

177. See Florian Hertweck, *Positions on Emancipation: Architecture Between Aesthetics and Politics*, Lars Müller Publishers / University of Luxembourg, Zurich: 2018.

178. See Architects Declare, UK Architects Declare Climate and Biodiversity Emergency, 30 May 2019. Assessed 11 August 2020. https://www.architects-declare.com

If theory became discredited by events and its own failures, we instead need effective manifestoes, which impel us into action — and into considering that the time to act is right now. Every appeal is necessary and welcome. All declarations must resonate as widely as possible, from the education field to the professional one, as well as through all available media. Many voices must rise and join groups such as *Architects Declare*, when they stressed that, even if "the research and technology exist for us to begin that transformation now", architects and clients still lack the "collective will" to "commission and design buildings, cities and infrastructures as indivisible components of a larger, constantly regenerating and self-sustaining system."[178]

And many other voices should join architecture critics such as Rowan Moore, when reporting the claims of Architects Declare on a relevant echo chamber, and pointedly asking "where are the architects that put the environment first?" As Moore relayed it at *The Guardian*, "architects (and for that matter contractors, clients, engineers and everyone else involved in making buildings) have no excuse for not giving their utmost to make their work have as little impact on the environment as possible." So, they must now "consider everything" — from "how far stone might have to travel from quarry to site" to "whether or not a building's components will end up as landfill when it is demolished." In a profession that supposedly "tends to attract people who want to change the world for the better," it is about time that design impulses are directed at climate action or, as Rowan brashly words it, at "the prevention of environmental and societal collapse." Indeed, the usual "squabbles about architectural style or form seem trivial by comparison."[179]

In *Ecological Architecture, A Critical History*, James Steele similarly converges towards the idea that, in "a world increasingly awake to environmental damage," the visionary eccentricities of the past are vindicated as "today's legal requirements," and every architect now bears "an obligation to the environment as well as to his or her client."[180] For a long while, the mandates of an uncharted ecological philosophy stood in apparent contrast with Modernist doctrine. Now, however, that opposition must swiftly dissolve. Rather, it would be a disgrace that, in a few years' time, we would look back at the architecture field and discover that the squabbles and the inaction had prevailed — even if calls for a practice transformation around the climate emergency and the implications of the Anthropocene have become loud and clear.

179. See Rowan Moore. "Where are the architects who will put the environment first?" 31 Aug 2019. The Guardian, Accessed 25 Feb 2020. <https://www.theguardian.com/artanddesign/2019/aug/31/architecture-to-counteract-climate-change-ilford-market-cork-house-barangaroo-mycelium>

180. See James Steele, Ecological Architecture, A Critical History, Thames& Hudson, London: 2005

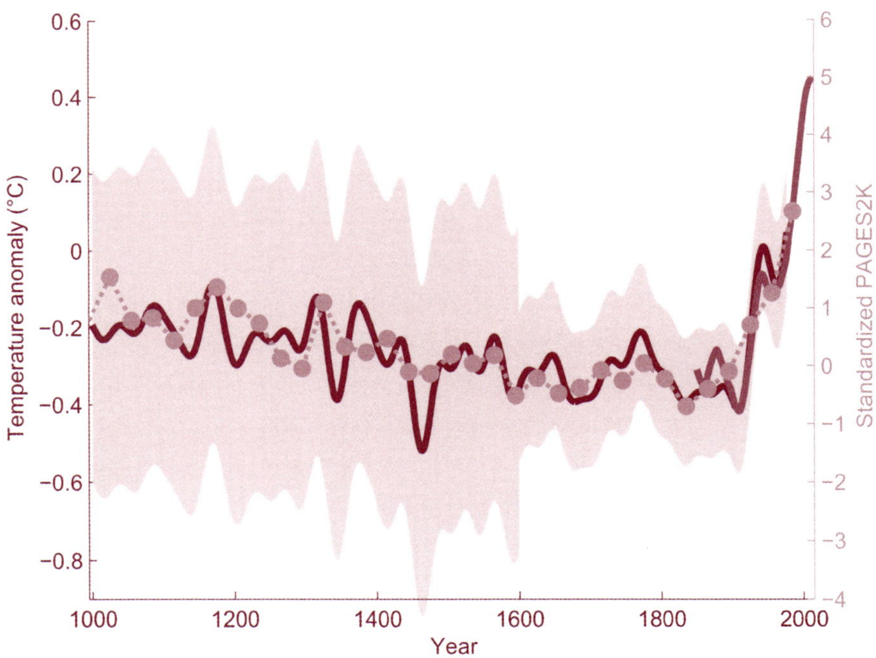

Hockey-stick Graph

In this respect, we should learn from the must-read account of *Losing Earth: The Decade We Almost Stopped Climate Change*. In it, New York Times journalist Nathaniel Rich described with alarming detail how a lost decade of climate discussions in the 1980s eventually led the world to a dreadful impasse.[181] It can be despairing to recognise that, after conditions arose for an early success in fighting climate change, it was actually its denial that gained thrust enough to stall any institutional and political decisions in that direction. Had action began then, we could now be at a very different stage of what today is still a would-be *Great Transition*. At this moment, instead, we must hope that in another three decades we will not be lamenting the time we have also lost circling the environmental crisis in the architecture realm, especially as soon as we realised that we actually owned the design and planning tools we needed to move into action.

As we also did here, researchers such as Damian White have already identified design pathways through which we can achieve a just transition into a more tenable ecological commons. As this sociologist sees it, that transition may first require social protests and forceful activisms. Yet, after the demonstrations and the eventual upheaval of the fossil-fuel *status quo*, "a just post-carbon world that actively improves people's lives still has to be imagined and built, fabricated and realized, institutionalized and sustained by public support and ongoing engagement." In this context, "a politicized mode of design" could provide the spaces in which the necessary dialogue between many disciplinary voices, technical skills and forms of knowledge could thrive — and ensure we can continue to "survive and flourish on a warming planet."[182]

As we accrue the dimension and urgency of the impending ecological catastrophe, we cannot but stress the urgency of the paradigm shift that is required in our built environment practices. Only in the week I'm writing, shocking reports confirmed that Australian fires killed an astounding 3 billion animals, i.e. nearly half the current human population.[183] With unprecedented forecasts of more than 25 storms in the United States, the

181. See Nathaniel Rich, *Losing Earth: The Decade We Almost Stopped Climate Change*, 1 August 2018. New York Times. Assessed 11 August 2020, https://www.nytimes.com/interactive/2018/08/01/magazine/climate-change-losing-earth.html

182. See Damian White, ibidem.

183. See Niall McCarthy, Australia's bushfires impacted over 3 billion animals, 4 August 2020. World Economic Forum. Assessed 11 August 2020, https://www.weforum.org/agenda/2020/08/3-billion-animals-were-impacted-by-australias-bushfires/

coming hurricane period was announced as "one of the most active seasons on record."[184] In the same days, these events were brought together with Covid-19 to warn on "the steady danger of overlapping disasters," and offer "a preview of life in a warming world."[185] Meanwhile, scientists did revise their estimates of global warming over the next decades, so that temperature rise is now more likely to fall into an interval of 2.6 to 4.1 degrees Celsius — "with a best estimate of slightly above 3°C." The same post from the World Economic Forum reiterates that "without actions to reduce emissions, [carbon dioxide] concentrations are likely to reach 560ppm — double pre-industrial levels — around the year 2060."[186] Yet, if we still do not grasp these events as affecting the balance of one integrated, delicate ecological system — or if, more likely, we are only psychologically overwhelmed by the extent of these disasters — such data may still not feel close enough to push us into action.

If the need for change is becoming undeniable even on the global political agenda, it is also necessary to show that such shift cannot be further delayed in the different realms of the built environment — and that, as revealed throughout this book, there are already conditions and capabilities in place for change to immediately begin. As some have claimed for the crucial arena of landscape design,[187] architects too must embrace a novel eco-political stance in sync with the inevitabilities of the Anthropocene condition. Either out of a genuine and knowledgeable concern, or out of convenience in keeping their leading positions, some global architecture companies have realized this urgency, and have already declared their commitments toward new behaviours.[188] Now, this mood must become generalized.

Beyond cautionary visions of "eco-gentrification and greenwashing,"[189] past immediate practices of adaptation and mitigation, and even after a much-needed "great carbon reckoning,"[190] the most poignant climax shift we can embrace in our field is probably the one that leads us to an architecture that contributes to ecological regeneration, rather than just adding to environmental damage. Further, as critics steadily start to

184. See Henry Fountain, "Hurricane Forecast: 'One of the Most Active Seasons on Record'," 6 August 2020. New York Times. Assessed 11 August 2020. https://www.nytimes.com/2020/08/06/climate/hurricanes-noaa-prediction.html

185. See Christopher Flavelle and Henry Fountain, "Hurricane, Fire, Covid-19: Disasters Expose the Hard Reality of Climate Change" 4 August 2020. New York Times. Assessed 11 August 2020. https://www.nytimes.com/2020/08/04/climate/hurricane-isaias-apple-fire-climate.html

186. See Piers Forster, Zeke Hausfather, Gabi Hegerl, Steven Sherwood, Kyle Armour, "Why low-end 'climate sensitivity' can now be ruled out," 30 July 2020. World Economic Forum. Assessed 11 August 2020, https://www.weforum.org/agenda/2020/07/climate-change-equilibrium-climate-sensitivity-global-warming/

187. See Bill Fleming, ibidem.

188. See Doug Voigt, "The rapid deterioration of Earth's biodiversity could be reversed equally quickly if we find new ways to design our cities." 29 August 2019. Dezeen, Assessed 14 August, 2020. https://www.dezeen.com/2019/08/29/climate-emergency-doug-voigt-som-opinion/

Solar panels made from food
waste by James Dyson

agree, "the aesthetic and material possibilities of a new ecological architecture could be exceptionally rich, driving an international artistic and political renaissance no less profound than the advent of modernism."[191] In contrast, although it may eventually be deemed necessary,[192] just fitting design and planning in the box of preparedeness — or the social, economic, and political emergency mechanisms that ready us for the worst — is unsatisfactory and insufficient. And finding refuge in the deceptions of a commercially commanded sustainability is no longer enough.

In her influential book, *After Geoengineering: Climate Tragedy, Repair and Restoration*, Holly Jean Buck seeks to transmit how we must "expand our thinking beyond mere sustainability," and definitely welcome the emerging notion of regenerative practices. This would mean moving from "doing less harm" or from just "letting nature recuperate" to actively working to increase flourishing." By this, she reinforces the idea that it is no longer enough to move "beyond an extractive, degenerate relationship with nature," but it is now necessary to engage in practices that heal and restore.

In passing, Buck also reminds us that Buckminster Fuller pioneered the notion of regeneration in the field of design. Already in the late 1960s, namely in *Operating Manual for Spaceship Earth* and its chapter on *The Regenerative Landscape*, Fuller clearly stated regenerative design principles as a prerequisite to elude a process of mass extinction driven by "over-specialization" and the *status quo* idea of "wealth." This direction, though, would require an invisible revolution of design and invention — which, most probably, and as abundantly illustrated here, only now do we have conditions to fully enact.

Buckminster Fuller was well ahead of his time in perceiving that "the fossil fuel deposits of our Spaceship Earth correspond to our automobile's storage battery which must be conserved to turn over our main engine's self-starter." Thereafter, he added, "the life regenerating processes, must operate exclusively on our vast daily energy income from the powers of wind, tide, water, and the

189. See Miriam McGarry, "Architects Declare a Climate Emergency But Can They Avoid Real Estate's Greenwashing Tendencies?" 13 Feb 2020. FailedArchitecture.com. Assessed 12 August, 2020. https://failedarchitecture.com/architects-declare-a-climate-emergency-but-can-they-avoid-real-estates-greenwashing-tendencies/

190. See Audrey Gray, "A Great Carbon Reckoning Comes to Architecture." 9 December 2019. Metropolis magazine. Assessed 14 August 2020. https://www.metropolismag.com/architecture/great-carbon-reckoning/

191. See Phineas Harper, "We're seeing an unprecedented mobilisation of architects in the fight against climate change." 11 June 2019. Dezeen. Assessed 14 August 2020. https://www.dezeen.com/2019/06/11/radical-architecture-climate-change-opinion-phineas-harper/

192. See Amy Frearson, "Architects must plan for "armageddon situations" says Snøhetta's Kjetil Trædal Thorsen." 12 December 2019. Dezeen. Assesses August 14 2020. https://www.dezeen.com/2019/12/12/snohetta-interview-kjetil-traedal-thorsen-carbon-negative-climate-change/

direct Sun radiation energy." Yet, it took us decades to make such renewable resources viable, as it will probably take us a few more to have them completely substitute carbon-based energy and ensure the *Great Transition*. For this transition to be successful we will need self-sufficient buildings, planned degrowth and ecological commons. But, as argued here, we will also need regenerative processes to become the heart of design orientations across the board — from zero-waste biotecture to innovative biomimicry design, and from interspecies-minded, biodiversity-fostering projects to the natural production of negative emissions throughout the built environment. And, given those pervading hockey stick charts, we will need all of these pathways as soon as possible.

After his seminal books on deep ecology or the ethics of the built environment,[193] philosopher Warwick Fox addressed such urgency in pretty clear terms in a paper from 2009. As he said it, "we simply have no option but to give overwhelming importance in the way we design, build, and plan to responsive cohesion with the biophysical realm." Already then, Fox spoke of a "transition architecture" or a "transition urban planning" necessarily taking over outpaced notions of sustainability. He went on to argue that, vis-à-vis science-based current scenarios of "a global collapse before the middle of this century," it would be "actually unethical" not to adopt a contextual design approach that, well beyond the basic needs of its internal cohesion, privileges the natural realm over the human social realm. "We are now all involved in a very high stakes game, whether we like it or not, and all previous, business-as-usual kinds of bets are off," Fox concluded more than a decade ago.[194]

Naomi Klein's *This Changes Everything* generally pointed in the same direction. For Klein, the climate crisis isn't just another "issue" that we can "add to the list of things to worry about next to health care and taxes." As she bluntly put it, this is "a civilizational wake-up call." It thus requires rather fast paradigm shifts in all areas of life: from our consumption behaviours to how we conceive objects, buildings and cities around us, from the way we operate economic systems to how we manage the built environment.

193. See Warwick Fox, *Toward a Transpersonal Ecology: Developing New Foundations for Environmentalism*, Shambhala Publications, Boston: 1990; *Ethics and the Built Environment*, Routledge, London: 2000; *A Theory of General Ethics: Human Relationships, Nature, and the Built Environment*, The MIT Press, Cambridge, Mass: 2006.

194. See Warwick Fox, "Ethics, Architecture, Responsive Cohesion, and the Transition to a More Habitable Future," paper presented to the "Ethics and the Built Environment 2009" conference, Nottingham University, 9-11 September 2009. Assessed 14 August 2020. https://www.warwickfox.com/files/2009-EB-Nott-conf.pdf

195. See Naomi Klein, ibidem.

The overlapping succession of everyday disasters affecting planetary ecosystems will increasingly reaffirm that "we need an entirely new economic model and a new way of sharing this planet."[195]

But Klein also reminds us that because of "decades of collective denial," it has become evident that "no gradual, incremental options are now available to us." With an unsustainable consumption-based economic system made global, we got to a point in which "gentle tweaks to the *status quo* stopped being a climate option." Indeed, for Klein, "our economic system and our planetary system are now at war." The environmental crisis asks for "a contraction in humanity's use of resources," while our current economic model demands precisely the opposite. And while radical change is now more than just a battle cry of radicals, the laws of nature would hardly be the ones being twisted in such war.

In this vein, it has been sarcastically claimed that it will be easier to destroy the planet than to disrupt the capitalist system. Critics have even brought this catchphrase into analyses of how hard is to envisage a growth-free future in the design world.[196] Yet, the recent global pandemic has suggested that the willingness to save lives could overcome the imperative to sustain the economic *status quo*. As the urgency of the environmental emergency becomes more and more undeniable, that tendency will hopefully prevail. As Klein again suggests, "surely humans should be capable of making some important lifestyle changes in the interest of stabilizing the physical systems upon which all of life depends." Even with the incentives of short-term benefits such as pollution reduction or new jobs in the renewable energy sectors, the problem now lies in when these changes effectively begin to take place.

When it comes to architectural practices embracing radical change as a means to tackle the climate crisis, it would rather seem like we can no longer afford to wait for the economic superstructure or lifestyle options to become more accommodating. The stakes involved in the built environment's contribution to ecological degradation are already too high. So, as at the outset of the modern movement, architects must take the lead in telling society: start building in a new way.

196. See See Federica Buzzi, ibidem.

As in one compelling tune by songwriter Josh Tillman, when it comes to start the change, one must necessarily conclude "why not me, why not you, why not now?"

IMAGES CREDITS

Page 24:
Marco Petrini / Marc Thorpe Design

Page 26:
Matías Riveros

Pages 28, 106:
Earthship Biotecture

Page 32:
Haus Rucker-Co

Pages 34, 39:
Ant Farm

Page 37:
Bruce Damonte / WORKac

Page 41:
Drew Angerer / Getty Images News

Page 46:
Superflux

Page 54:
Talia Woodin

Page 55:
UK Architects Declare

Page 58:
The McHarg Center and Billy Fleming

Page 63:
Adrian Lahoud

Page 66:
Pius Utomi Ekpei / AFP

Pages 70, 71:
Forensic Architecture

Page 76:
Building Transparency

Page 83:
Studio Generate

Page 84:
KPMB architects

Page 88:
Rasmus Hjortshoj / BIG-BJARKE INGELS GROUP

Page 94:
Anyone Corporation

Page 99:
SUPERFLEX

Page 103:
Andrés Jaque / Office for Political Innovation

Page 108-109:
The Blink Fish / Boeri Studio

Page 113:
Senate Democrats

Pages 116, 150-151:
SCAPE

Page 121:
Actar Publishers

Page 128:
METROPOLIS Magazine

Page 131:
Druot, Lacaton & Vassal

Page 132:
Superuse Studios

Page 134:
The Architecture Lobby

Page 143:
ISOHEMP

Page 144:
Daniel Shearing / Waugh Thistleton Architects Ltd.

Page 149:
BIG-BJARKE INGELS GROUP

Page 160:
OAT / Rosa Nussbaum

Page 161:
NORELL/RODHE

Page 162:
Kate Yeh Chiu and François Sabourin

Page 163:
OAT / Istvan Virag

Pages 170-171:
Iwan Baan

Page 182:
Annar Bjørgli / Nasjonalmuseet

Page 185:
Julien Lanoo

Page 186:
Rohspace / UNStudio

Page 191:
Allard van der Hock / Superuse Studios

Page 193:
ICD-ITKE

Page 196:
Images by NLÉ

Page 198:
Studio Symbiosis

Page 199:
Framlab

Page 201:
Nicole Moyo

Page 202:
Giovanni Vaccarini Architects

Page 207:
Dana Cupkova

Page 208:
Exploration Architecture

Page 210:
Stad in de Maak

Page 217:
Bruno Lopes / Maat

Page 218:
Rimini Protokoll / Royal Academy of Arts

Page 227:
AuREUS

CLIMAX CHANGE!
How Architecture Must Transform in the Age of Ecological Emergency

AUTHOR
Pedro Gadanho

PUBLISHED BY
Actar Publishers
New York, Barcelona
actar.com

COPY-EDITING
Angela Kay Bunning

GRAPHIC DESIGN
Ramon Prat Homs

Distribution:
Actar Distribution Inc.

New York
440 Park Avenue South,
17th Floor
New York, NY 10016, USA
T +1 2129662207
salesnewyork@actar-d.com

Barcelona
Roca i Batlle 2-4
08023 Barcelona, Spain
T +34 933 282 183
eurosales@actar-d.com

ISBN: 978-1-94876-567-1
PCN: Library of Congress Control Number: 2020946946
Printed in Europe, 2022

"Cimax Change!" is a research project funded by the Graham Foundation.
grahamfoundation.org